"It doesn't take long for the book's hard, black lines of type to disappear, and for the reader to be spellbound, completely submerged in Simonds' special world. . . . While winter and spring play tug-of-war and your snow shovel still sits next to your gardening tools, here is this great book to help you settle your yearning to poke your fingers in the dirt and plant your early seeds of spring." —THE GLOBE AND MAIL

Praise for **A NEW LEAF**

"A vibrant, exuberant hybrid—part memoir, part literary gardening essay." —*More Magazine*

"Even gardeners with limited space will find much to delight them in this book. So will complete non-gardeners: like an armchair bullghter or ballet dancer, you don't have to do it to enjoy reading about it." —*The Gazette* (Montreal)

"Two green thumbs up for Merilyn Simonds. This shining book, sparked by her garden blogs on everything from bugs to peas to survivor elms, evolves as a gardener does, from hope to realism and back to hope. Simonds reminds us that our contact with each garden bed (she has twenty-six, one for each letter of the alphabet!) lets us touch yet another aspect of our selves. *A New Leaf* is a by-your-bedside companion, composed to seed our dreams."—Molly Peacock, author of *The Paper Garden*

"Like stories passed friend to friend, these wise, funny, colourful pieces enrich our understanding of plants, landscapes and life. A book to grow by, and share." —Sarah Harmer, singer-songwriter

"Delightful, funny, wise. . . . In the tradition of the best gardening books, Merilyn Simonds' A New Leaf inspires both experienced gardeners and those just beginning."—Beth Powning, author of *The Sea Captain's Wife*

"I certify Merilyn Simonds the Saint of Frugal Gardening, for her amazing and helpful skills with plants, other edibles, and people too."—from *The Year of the Flood*, God's Gardener Scroll, awarded by Margaret Atwood

A NEW LEAF

GROWING WITH MY GARDEN

Merilyn Simonds

ANCHOR CANADA

Library and Archives Canada Cataloguing in Publication is available upon request.

Issued also in electronic format.
ISBN 978-0-385-67047-0

Printed and bound in the USA

Book design: Leah Springate
Cover photos: (front) Tohoku Color Agency / Getty Images; (tip-in)tk;
        (back) © iStockphoto.com / Argument
Author photo: Erik Mohr

Published in Canada by Anchor Canada,
a division of Random House of Canada Limited

Visit Random House of Canada Limited's website: www.randomhouse.ca

10 9 8 7 6 5 4 3 2 1

For Astrid & Estelle

A GARDENER'S CREDO

I AM NOT WHAT SOME WOULD CALL a serious gardener. I don't know the Latin names of plants, except those that sound subversive or whimsical or mysterious. *Phlox subulata. Euphorbia corollata. Nepeta nervosa.* I try to design my gardens by the book—three of this, seven of that, never four or six—but in the end, I do what looks good to me, because let's face it, no bus tour will ever traipse across my white-clover lawn. I will never show my delphiniums at the fair.

My Beloved laughs when I say I'm a lazy gardener. It's true that I'm out the door at dawn and he has to drag me back inside when the sun goes down. But I don't plant my carrots in rows or deadhead my dahlias, and I never (almost never) turn the soil. I rarely water. Only the babies in my garden beds are coddled.

I have land, more than enough land, but not much money and less time. I don't want to work any harder than I have to.

Pleasure is the only rule. The exuberant sweep of colour, the sweet scents and sharp tastes, the upthrust trailing shapes, the accidental pairings that make me laugh or weep with their unlikely beauty—we're bound together, my garden and me, in an ecstasy of growth.

CONTENTS

FOREWORD

FOR MOST OF MY LIFE, I've grown the food my family eats and the flowers that bring beauty to our table. I've often thought, as the world in which I live grew affluent and the drift toward urbanization became a tsunami, that the skills I have accumulated would wither with me. No one needed to know how to cure garlic or when to harvest beans or the best time of year to prune an apple tree, how to make jam without adding pectin or cook a compost pile. Certainly no one was interested in how to make a chicken come when it's called or which flowers can be eaten, which will cure, and which can kill. Once, this was essential knowledge. Not anymore, I thought.

But the wheel has turned and here we are again, wanting slow food, uncontaminated and organically grown local food, food that we can trace to its home soil. Flowers without that bitter florist scent. Blooms we can eat and drink and float in the tub and savour every minute of their brief lives, and ours.

And something else, too. We hardly know how to express it, it's such a deep and diffuse yearning, like an ache with no clear cause, though we know when it is soothed. The same urge made our childish selves splash in puddles and fashion caves in the woods—an urge that is satisfied by peeling back the grass and laying a hand on the warm and living earth.

This book is the story of my gardens at The Leaf. At the turn of the millennium, my Beloved and I bought a two-hundred-year-old stone house situated in what was left of an old orchard after the Great Ice Storm of '98. We opened the soil for vegetable beds, fruit beds, tea beds, herb beds, perennial beds, a Woodland Garden, a garden of ephemerals, another for native plants, and a Hortus Familia where I grow species that honour our mothers and fathers and where we bury our pets. In all, twenty-six beds. An alphabet of plants. That's it, I told my Beloved. When I feel the urge for another, I'll write about it instead.

On March 21, 2009, the first day of spring, I launched frugalistagardener.com, a website where every week I post an essay prompted by my gardens. These are not instructional pieces, although a discerning reader might pick up hints on pre-sprouting beans and splitting hostas. Reading them is more like spending an hour wandering the garden paths with me, kneeling in the beds, crushing a slug, pushing a hand into the soil, marvelling at what is there.

What people respond to on the website, and in this book, too, I hope, apart from the joy we share in the presence of things botanical, are the stories, the characters: the Rosarian next door, who calmly teaches; the Frisian, who comes once a week to weed; the Garden Guru, who guides the evolution of the beds; my Beloved, who muses laconically on all that we do.

These short personal essays are intimate, meditative, and humorous, filled with wonder and a questioning eye. They evolve through the course of a gardening year, moving backward and forward in time, from the making of this garden to that of every garden I've ever worked, meandering into some of the great gardens of the world, coming back always to the soil within reach, to the pleasures and frustrations that force me to grow in my garden, too. For who can watch the brief cycle of a pea without contemplating one's own life trajectory?

Through it all is woven the motto that guides my hand: never work harder than you have to; live as gloriously as you can.

FRESH GROUND

INTO THE PLOT

ALL WINTER THE GARDEN WAS LIKE A CLOSED-UP RESORT, rooms echoing and vacant, white fabric draped over the furnishings. Now the sheets have been yanked off—the snow melted that fast—and already the regulars are coming back.

The crows arrive first. They come as a couple, though they don't stick together. One pokes along the edge of the woods, nosing in the verge, while the other struts across the grass like a maître d' inspecting the premises. Maybe they take turns, one strutting, one checking out the woods for nesting sites: I don't pretend to be able to tell them apart. Both are big as ravens, and glossy, their beaks held haughtily in the air.

The vultures aren't far behind. They skim the canopy, circling our yard, sniffing for the bodies of winter-killed rodents uncovered by the shrinking snow. The bare-skulled birds often land in the trees at the rim of the woods, but never close to the house. "We aren't that old yet," my Beloved declares.

Then suddenly, the rest are here. Red-winged blackbirds from the field across the road swarm the feeder on icy mornings and on those days when March sends a sleeting white reminder that winter's not over yet. Flickers bob under the apple trees, pecking for crumbs in the grass. The goldfinches begin a slow striptease, throwing off their dowdy winter duds for summer bling. But it's the Canada geese we wait for, the ones that spell spring with lines in the sky. They flock by the thousands to the Farmer's cornfield across the way, exhausted from their journey north across the lake. All night they honk and chatter as if they can't wait until morning to share stories of their travels.

For days, sometimes weeks, the geese are the soundtrack to my garden cleanup, a discordant, percussive jazz that goes on into dusk, with themes that recur, cadences that rise and fall as if there might be an intent to it after all. In the foreground, the chickadees and song sparrows, finches and blue jays and cardinals slip into their courting songs, and before long, the phoebe is back, screaming, "*Phoebe! Phoebe!*" and the wrens are whistling their sweet melodies and the catbird is imitating everybody, even the scrape of the saw as my Beloved prunes the apple trees. Chipmunks scoot along the stone wall, scooping up the dangling seed heads before I cut down last year's stems and bury them in the compost. Squirrels chase one another in a mating marathon under the drooping dogwood and up the ornamental cherry, leaping to the locust, then to the sugarplum trees, racing pell-mell into lust.

It's a party out there and I'm not invited. No one is. We're all crashers on this first day of spring. I lather my winter-softened hands with Bag Balm, pull on my rose-covered shirt and my new green garden gloves, and step out to join the rave.

HEARTWOOD

Standing with my back to the old split-rail fence that defines the western edge of our property, I can make out four rows of apple trees—not the deliberately stunted specimens that have taken over modern orchards, but big old apples, with trunks too thick to embrace. The limbs start low and spread generously, inviting a shinny up to sit splay-legged over a branch, gazing down on the blossoms and birds. Trees as open-hearted and sheltering as great-aunts.

It was spring when we moved to The Leaf, a parcel of forest and meadow along a stretch of country road in eastern Ontario where squat stone houses declare a fiercely humble intention to stay. We knew no one here—that was part of the appeal. The snow was just retreating from around the stumps of the trees felled by the big ice storm two years earlier. Half a dozen of the old apple trees were cleaved down their centres, branches torn off by the weight of water, exposing heartwood the colour of wounded flesh.

We had a vision of sinking our teeth into midwinter apples, crisp from the cellar. Of toasting each other with cider pressed from culls raked off the grass. And so the day after a mighty flock of Bohemian waxwings stripped the last of the thawing, fermenting fruit from the limbs, we set about restoring what remained of the orchard.

My Beloved and I assembled ladders, chainsaws, pruning shears of various sizes and enlisted the help of our friends, the Carpenter and the Garden Guru, who had done this before.

"A bird should be able to fly freely through the tree when we're done," they said.

We started with the apple tree closest to the house, the one I could see from the kitchen window. The Garden Guru walked around it slowly, eyeing it like a piece of marble she planned to chisel. My Beloved and the Carpenter positioned the ladders and revved the chainsaws.

"That one," she said, and the Carpenter squinted to find an outside bud, then trimmed the limb close, so the next branch would grow outward instead of toward the trunk.

Bring the height down, open the centre to the light, balance the spread of limbs, she recited as she circled the tree. Before she called out each cut, I tried to guess which branch she'd choose, and why. Wrong. Wrong. And then, suddenly, I got it right.

We did four trees that afternoon, carrying armloads of pruned branches into the house, where they burst into blossom by the fireplace, a foretaste of true spring. My Beloved and I pruned three more trees on our own, uncertain in our cuts, anxious not to remove more than the 25 percent per year the trees could withstand without going into an arboreal version of cardiac arrest.

It wasn't until early May that we realized the extent of the original orchard at The Leaf. Only sixteen trees remained, but if we followed one burst of white bloom to the next, as if in a game of connect-the-dots, we could see where rows of apples had once run the entire width of the property, interrupted only by the house. The row closest to the road bore fruit early in August; the third produced later that month. The fourth, all but lost in the fringe of sumac and saplings at the edge of the woods, were Russets, we could tell by the brown mottled skin. The trees in the second row, the ones I could see from the kitchen, produced apples that we swore were McIntosh.

And they were. One summer afternoon in our third year at The Leaf, an elderly couple stopped their car at our mailbox. There's little traffic on this road except for the school bus, the milk truck, the snow plow, and an occasional commuter or lost tourist. We weren't used to company.

"I'm Apple Annie," the woman said. She was born in our house, seventy years before. Her father planted the orchard in 1923, ninety trees set in four neat rows.

"Jonathan, McIntosh, Scarlet Pippin, Russet," she said, naming the rows. Her parents stored the apples in barrels and shipped them off to local hotels and grocery stores. As a girl, she sat by the road and sold them to passersby. "That's how I got my nickname," she laughed. When she grew up, she went to university to become a teacher, then lived at home, riding her pony through the woods to a one-room schoolhouse up on Washburn Road.

I ran to the stone wall to retrieve a horseshoe my Beloved had unearthed as he dug another garden, and handed it to her.

"In the fall," she said, turning the shoe upright to hold in the luck, "the whole place smelled of apples."

It's the same in the summer, when the young, green fruit give off a spicy scent that fills the West Yard. Sometimes, my

Beloved and I position a small table under an arching apple branch and linger there after our lunch, books in hand.

Once, the Farmer who works all the land that we can see from The Leaf stopped his tractor beyond the hedge that separates our yard from the road. The afternoon was wearing on, the sun already sliding down the sky, and we were still sitting with our books, stirring briefly whenever the hay wagons passed, feeling vaguely guilty in our indolence. I couldn't help but recall what our nearest neighbours, the Rosarian and the Humanist, had told us—that the Farmer had once interrupted their long afternoon of reading to climb down off his tractor and ask, "What are you doing anyway, sitting there all this time?"

Now the Farmer was striding through a break in our hedge. Small and wiry, he still works dawn to dusk beside his sons, though he is well over eighty. "Never tasted a drop of alcohol," he said the day he stopped to welcome us to The Leaf. When we offered him a cup of tea, he added: "Never took a hot drink, neither."

We could see his lips moving as he strode across the grass, and we expected the worst. Then we heard him, a clear, true tenor, *Don't sit under the apple tree with anyone else but me . . .*

"I've been watching you all afternoon, and singing this song. I just wanted to thank you for that," he said, beaming broadly. Then he turned and left.

Every March since, we pull out the chainsaw, the pole pruner, and the secateurs and give the old trees a trim. As I haul away the last armload of this year's trimmings, a robin glides through the branches with outstretched wings. I think of Apple Annie and her father, who set out the tender scions; of our friends, who made the first cuts; of the Rosarian and the

Humanist, and the Farmer who sang to us. We bought this place with notions of solitude, but already, there is a gathering on the lawn under the old apple tree, and I feel at home.

AS GOOD AS YOUR TOOLS

As a child, I loved tools. My mother's wire cake beaters that popped out of the machine at the press of a button. My father's brace and bit with the knobby wooden handle worn shiny from use. Because they could do serious harm, my father's tools were more compelling, drawing me down to his basement workshop when I heard the pound and scrape of hammer and saw. He'd let me stand beside him on an old egg crate and hand him squares of rough red, grey, and orange sandpaper. Before I could read or write, I knew the difference between 80- and 120-grit. I loved the way the numbers were printed all over the back so you could never make a mistake.

Once, my father helped me build a closet for my dolls' clothes. He sawed the boards and let me fasten them together, then he fit the clothes rod into a notch he'd carved with a small, curved chisel brought to a fine edge with a grey slip stone. I wasn't allowed to touch the sharpening stones. They were lined up out of reach, meticulous in their rack along the back of the workbench. Oil stones and water stones, slip stones and dry stones, one for every purpose: the knife that carved the Sunday roast, the scythe that cut the long grass at the bottom of our yard, the chisel that gouged the seat for my closet rod.

"You're only ever as good as your tools," my father said.

I have my own tools now. They lie in a heap in a cupboard outside the kitchen door. Secateurs and edgers, clippers, loppers, hand-hoes and trowels, shears, scissors, and saws of various sizes. I know enough to buy good tools and a good black dress—legacies from both my parents—but whereas the dress hangs in a zipped plastic bag at the back of my closet, my tools are mud-smeared and rusted, their blades dull from use.

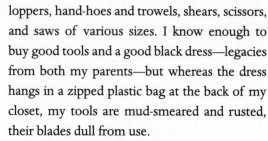

I know enough to buy good tools and a good black dress—legacies from both my parents.

"I'd rather use my hands," I say to my Beloved. It is an ongoing argument. He reaches for a tool at the slightest provocation, whereas I love the feel of plants and soil, would rather pull a weed than dig it, rather snap a stem than cut it. My childish affection for tools has been tempered by experience. Tools are expensive. They get in the way. "I am," I say with stubborn pride, "nothing like my father."

A few years ago, our friends the Carpenter and the Garden Guru came to stay at The Leaf when their apartment was gutted in a fire. They brought a trunkful of tools that they laid out on the grass in the spring sun. Carefully, they wiped each tool clean of creosote and soot, honed each blade, and oiled the wooden handles until each one shone, not like new, but like a thing made precious by fond use.

That should have been my burning bush. But it was the Rosarian down the road who finally made me change my sinful ways. A retired professor, the Rosarian has kept that fussy, academic attention to detail and deep respect for the past that are the marks of a true historian.

Every morning, before he enters his plot of 256 roses, he sits at his kitchen table and hones his secateurs on a small grey stone. When the blade is sharp enough to slice a hairline across his thumb, he rubs the metal down with an ancient oily cloth, then he wraps the cloth around the sharpening stone and the oil can and sets them in a small wooden picking basket that he keeps by his chair.

I had come upon him doing this a dozen times or more, on those mornings I dropped by with fresh eggs or when I came looking for his famous recipe for feeding roses, which I misplace every year.

"I should do that, too," I'd say, watching him draw the blade across the grey surface in a single sweep, as if peeling the stone.

"You should," he'd agree, knowing that I wouldn't.

And I never did get into the sharpening habit, not until this spring, when he came to help me prune the roses. I've never liked roses, but there were two spindly shrubs here when we bought the place and I feel obliged to care for them. They were in the way of a stone wall my Beloved was building, so I moved them, carelessly heeling them into a hastily dug hole on the far side of the Kitchen Garden. They survived. The leaves were spotted, the foliage sparse, and the thorns an argument unto themselves, but the blooms had a fragrance like nothing else on earth, so I let them stay. I fed them. I even watered them on occasion. This spring, I asked the Rosarian to show me how to cut them down to size.

"See the brown at the tip? Make the cut where it turns to green." He's a good teacher. He refuses to make the cut for me. He watches with a critical eye as I bend to the stem, clamp it between my tarry blades, and squeeze.

"My hands must be getting weak," I say, grasping my rusted secateurs more tightly and giving them a firmer clench. The blades bury themselves in the mashed fibres of the twig.

"Try these." The Rosarian slides his shiny cutters out of the leather holster looped on his belt, ruining me forever.

Spring is coming on quickly. I scurry to find the small grey honing stone I took from my father's workbench before we sold his house, a can of lubricant, a square of purple flannel from the winter nighties I made for the Grand Girls, daughters of my Younger Son. I hone my good Felco secateurs first, then a cheap pair I bought at the local hardware and abandoned when I could no longer muster the strength for a proper cut. I hone the loppers. I hone the shears. I put a sharp edge on the shovel. The trowel. My favourite loop-hoe. The dividing spade. The hedge trimmers. The grass cutters. My Beloved stops me when he catches me prowling the shed, eyeing the pitchforks.

I vow to do this every day before I take my tools outside. I swear it on a stack of William Dam seed catalogues. The slick slap of metal across the honing stone will become my morning prayer, the resonating *om* that starts my gardening day.

Already, this honing has worked miracles. My fingers, only yesterday too weak to make a decent cut, are made strong overnight. I can lop a lilac sucker the size of a sapling, snip errant branches for hours on end. Suddenly, I'm a strongman, a Hercules. More Samson than Delilah. I am, at last, my father's daughter.

ONE GOOD TURN

WHAT IS IT, THIS URGE TO DIG, to peel back the earth's skin and sink my fingers into moist soil, to manipulate what grows there of its own accord?

It is not a simple thing.

We arrived at The Leaf in the spring of the millennium. Fifteen acres of woods with a strip of mown meadow around the house, along the road. From the kitchen, looking west to the two-hundred-year-old boundary fence of split rails, there was only grass, a bank of cedars that divided the space roughly in half, and an apple tree off-centre in the nearer distance.

I saw a room. A banquette of flowers around the apple trunk. A shrubbery to wall off the road, and on the opposite side of the yard, a rise of perennials, for balance. At the far end, a border of *Rosa rugosa* broken by two soaring limes, a pillared portal to the *allée* that would lead across the meadow, up into the woods.

A row of ancient peonies was heading the wrong way. I dug up the roots while the shoots were still ruddy pencil points and set them on the perpendicular. A lilac stuck in the middle of the yard was relocated to the Shrubbery, where it faltered for years before finally giving up. I scarcely mourned its passing. I hadn't planted it; it wasn't mine.

As soon as the dampness left the soil that first spring, I marked out a perennial bed, a hundred feet long by eight feet wide, and stripped off the sod. I grabbed a handful of earth to augur my gardening future. The soil in my northern garden had been pure sand. My city garden, which backed into a limestone bluff, had been pure clay. The Leaf marks the edge of the same great prehistoric sea that once flowed down over the plain where the city rises now, but I was hoping that the glaciers had deposited a scraping of loam on top of the compressed skeletons of sea creatures. I rubbed the soil between my fingers, squeezed it in my fist. There was almost no grit. Not sandy, then. The ball of soil was not loose. Not loam, then, either. I squeezed it into a thin sausage. Clay loam wouldn't be so bad, but the sausage bent without breaking. Pure clay.

I should have dug it out and carted it off, replaced it with free-draining soil, but I was puffed with the power of my vision. Clay was good, I told myself. Rich in minerals. It only needed its bonds broken for all that goodness to be released to the plants.

Organic matter, that would do the trick. My Beloved was busy moving stones for a wall he was building around the terrace, so I made a deal with the Boy Next Door. I would edit his high-school essays if he would help me in the garden. All through March and April, he carted manure and hay, layering it over the clay. When he offered to dig it in, I protested, "No digging! Let the worms do their work."

That first year, I chose plants that preferred the soil I had to offer. Plants that wouldn't mind the lack of drainage, the sticky gumbo of spring, the concrete that my dessicated clay would become in midsummer. Lupin. Yarrow. Delphinium. Aster. Shasta daisy. Cranesbill geranium. The lovely blue anchusa. Sea holly. A thistle that set coppery blooms like Kurly Kates.

Every summer I top-dressed with compost. Every fall I heaped the bed with leaves. Every spring the Boy Next Door wheeled barrowsful of manure-soaked hay from the chicken coop. He grew into a man and graduated, moved away to teach in a school and start a garden of his own. I hired the Frisian, a Dutchman from the Netherlands province of Friesland, where hard work, especially outdoors, is held to be a pleasure.

"It's time to double-dig," I said to the Frisian last spring.

I had come to this conclusion with reluctance. I believe in the soul of soil, the logic of worms. The Leaf is a no-till zone. Turning a garden bed is like going into a stranger's house and moving the freezer from the basement into the attic, stuffing the sofa and coffee table into an upstairs bedroom, the dressers into the main-floor powder room. How will the soil ever sort itself back into its rightful layers? Then again, how will my heaps of compost and mulch ever penetrate deep enough unless the soil is turned?

"Double-dig?" the Frisian said brightly. "We can do that!"

He proposed a marathon. One Friday in April, I dug up every plant in the west half of the long perennial bed, setting them in the big black nursery pots I'd collected over the years. Species that hadn't yet poked up through the soil I labelled, scrawling the names on slices of yogourt containers. At dusk, I was only halfway through the job.

The Frisian arrived Saturday morning at first light. He spread a tarp at one end of the perennial bed, and at the other, dug a trench, two feet deep and two feet wide. He deposited the soil on the tarp, got a wheelbarrow-load of manure and compost and dumped it in the trench, then dug another trench beside it, emptying the soil from the second trench on top of the manure and compost. I scrambled behind him, removing the last plants from his path as fast as I could. The deep-rooted anchusa I left where they were: they had obviously come to an accommodation with the clay. The lupins, too. They hate being moved.

By noon, the Frisian had leapfrogged his way down half the bed. By dusk, it was finished, the soil from the first trench neatly scraped off the tarp and heaped into the last trench of the bed. After he left, I stood in the sharply slanting light and wondered what it was we had done. Put nature's slow enriching of the soil on fast-forward? A good thing, then. Or had we disrupted something we didn't understand? Were there clay-loving micro-organisms gasping their last because I wanted a different kind of soil, right there, right then ? Was there some rare nematode, some all-but-extinct beetle that I had just deprived of a home, a last meal? I imagined worms turning in slow circles, dazed and confused after the apocalypse, their lives in ruins.

I think of them again today as I contemplate the eastern half of the bed. It needs remaking, too. In July, the clay shrinks, opening Death Valley crevasses at the foot of the scarlet bee balm. The golden marguerites grow lopsided and topple, as if in dismay. I look to the bed the Frisian double-dug last year. The plants fairly shout with joy. They have room to breathe.

I plunge my hand into the twice-turned soil. It sinks deep. The earth in my palm squeezes to a ball, a loose ball I can't shape to a sausage that curls, no matter how hard I try. A worm wriggles to the surface, then plunges out of sight again, a slow-motion diver. Or maybe that twisting and turning is a kind of dance. Darwin wondered if earthworms respond to music. He once moved his piano and various instruments out to the garden, and while his wife and children played minuets, he bent

close to the earth, observing the worms. I don't know what conclusion he came to.

Tomorrow I will be turning this worm-world upside down.
I hate the thought of it, but I don't know any other way to make
the soil good for my plants. I could leave it be, forget the gardens
altogether. Would the worms be happier without all that tasty
manure and mulch? Maybe they grow philosophical, too, as
they set their house in order, singing some nematode version
of that old Byrds song "Turn Turn Turn," which borrows from
Ecclesiastes: *To every thing there is a season, and a time to every
purpose under heaven . . .*

DANCES WITH DAFFODILS

EVERY APRIL MY MOTHER WOULD PIN a yellow plastic daffodil to her lapel and walk up and down the streets of our small village, collecting donations for the Cancer Society. I think of her often, but especially on the day my first 'King Alfred' opens. The petals are so perfectly formed, so rigidly glossy, that I exclaim to my Beloved, "They look just like plastic!" I mean it as a compliment.

'King Alfred' is my favourite among the daffodils. It's what is known as a trumpet daffodil: a tall, deep yellow narcissus with a trumpet longer than its petals. Hybridized in England in the 1890s, 'King Alfred' was named for the greatest of the British kings because it was considered (and still is, by me) the greatest of the daffodils. Alas, a true 'King Alfred' is next to impossible to find in nurseries today. The name has become generic, like Kleenex and Coke, code for "tall, deep yellow, long-trumpeted."

Daffodils are, to my mind, the very best of the spring bulbs. They don't ask for much more than a bit of April sun and rain to rise golden into the air. They multiply like mad, and because they are poisonous, squirrels and deer leave them alone.

The first spring we moved to The Leaf, my Younger Son gave me a hundred daffodil bulbs. I planted them in twenty nests of five. After a few years, I dug up one of the nests: five had become thirty-five.

What can a gardener do but sink to her knees before such generosity?

The trick is to figure out when to dig up a nest, separate the bulbs, and transplant. Once the daffodils are up, I can't bear to disturb them. After they have flowered seems no better: I don't want to interrupt the storage of nutrients that is vital to producing a good bloom the following year. But if I wait until that process is complete, the leaves will have rotted away and I won't find the bloody bulbs at all.

Years of doing nothing as I watched this cycle pass has made me swear to dig up the daffodils the minute they poke through the soil. I do this gently, separating the bulbs but keeping as much earth attached to the roots as I can. The new holes are already dug: I plop the bulbs in and cover them up before they have a chance to wilt.

Once, I forgot to dig the hole in advance. Just as I was lifting a nest, the heavens opened and let loose a downpour. I threw the clump of bulbs onto the ground at the front of the Shrubbery, where I'd intended to plant them, and kicked some grass-clipping mulch overtop for protection. In the chaos that is spring, I forgot about them entirely. They bloomed and faded, then lay there unnoticed under their scrim of clippings until the following spring, when they bloomed on top of the soil in a riotous bouquet.

Usually, though, I plant my daffodils deep, setting the base of the bulb six inches below the surface, at least. I don't mind that this makes them a little late showing their sunny yellow faces. Planting deep leaves me lots of room to set summer annuals on top once the bloom has passed.

When I first planted my daffodils, I set them boldly out in the sun, in beds all their own. I imagined a belt of daffodils such as Dorothy Wordsworth described after a walk she took with her brother William on a windy April day. High in the hills of the Lake District, they came upon a swath of wild daffodils as wide as a country road. "I never saw daffodils so beautiful," Dorothy wrote in *The Grasmere Journal*. "They grew among the mossy stones . . . some rested their heads upon these stones as on a pillow for weariness and the rest tossed and reeled and danced and seemed as if they verily laughed with the wind that blew upon them over the lake."

For her brother William, the encounter inspired a poem that every British school child—and my Beloved—can recite:

> I wandered lonely as a cloud
> That floats on high o'er vales and hills,
> When all at once I saw a crowd,
> A host of golden daffodils; . . .

I loved my host of daffodils, but what to do with the spike of leaves when the flowering is done?

"You could cut them back," the Garden Guru said without conviction.

What, then, would feed the bulb? I want the leaves, but I want them out of sight. So I have resorted to subterfuge. Every year, I dig a few more daffodils out of their showcase swath and plant them among the shrubbery and late perennials. The new growth springs up just as the daffodils begin to fade, camouflaging the wilting leaves.

In those places where I want the daffodils front and centre —in the foundation beds, for instance, against the backdrop of yellow cedar—I bend the browning leaves around my fist and tie them in a loose knot. I wish I could remember who taught me this. Every spring I rack my brain for the answer as I move among the beds, knotting daffodil leaves into nameless little topiaries.

It is scarcely April now, and I am waiting for those Wordsworth blooms. Not just the golden 'King Alfreds.' I have other daffodils, too: white daffodils, white with pale peach or lemonade trumpets, miniature daffodils. Some of these were gifts, some I got from friends who were thinning out a patch, and some I bought at the grocery store, an unexpected gold-mine of spring bulbs. Late in February and throughout March, grocery stores sell potted bulbs to winter-weary customers, but they inevitably overestimate the need. The leftovers are put on the mark-down shelf, where I find them, long past blooming, five or six to a pot for a dollar. I put them in a cool dark place, let them dry out, and plant them late in the spring, thinking of my mother and all those years I wore a daffodil pinned to my lapel for Mother's Day Sunday service, a coloured bloom to show my mother was still alive. Now I gather white daffodils for a Mother's Day bouquet.

My Friend in the North plants nothing but 'King Alfred' daffodils, in memory of her husband, the Beekeeper, who died ironically and tragically from the sting of a bee. She plants the bulbs in a meadow behind her house, where they have multi-plied and spread.

"My Asphodel Meadow," she says. The word *daffodil*, she tells me, derives from Asphodel, which is that section of the Greek underworld reserved for ordinary souls. The good and heroic spend eternity in the glorious Elysian Fields, while the very bad go to Tartarus, the Greek version of hell. The rest of us, who are sometimes very good and sometimes not, live out our days in Asphodel. I think of my mother there, with the Beekeeper, and maybe Wordsworth, too, wandering lonely as a cloud, and Dorothy, her new friend, their hearts filled with pleasure as they dance among the daffodils.

OLD-LADY FLOWERS

MY MOTHER WASN'T MUCH OF GARDENER. She kept a yellow potentilla blooming desultorily at the corner of the garage through the summer, planted a few tulip bulbs in the fall, and every spring set out a stiff row of geraniums, which lived like trolls in our basement through the winter.

I think of her every year as I snip the growing tips off the tender geraniums I bring indoors in October before the first killing frost. I don't do much of this sort of coddling. The rosemary bush, the bay laurel, the oleander and mandevilla vine, these unhardy geraniums: it's a small and select party that is invited to spend the winter inside the house at The Leaf.

My grandmother used to shake the soil off her fall-pulled plants and hang them upside down in the cellar, soaking the roots for a few hours every couple of months, then sticking them in the ground again come spring. My mother potted her geraniums and wintered them in the basement, in front of a narrow window in what we called the wreck room. I keep mine in the unheated and dimly lit guest bedroom, where they grow leggy and pale.

I'm a poor host. I keep the door closed and offer a drink only when I happen to think of it. The plants are resting, after all. As the light grows stronger, I trim them back, harvesting a few dozen cuttings for the coming year.

Looking down at my hands as I set the tender tips in pots and press soil against their stems, I see my mother's hands, the too-short fingers, joints grown gnarly with work. She ruined her hands mixing cement for the house she and my father built. I am ruining mine in the garden. She used to lay my slender young hands on hers as she pressed the ivory keys of the piano, trying to give me a feel for music. I do the same with the Grand Girls, cradling their fingers around a seedling, showing them how to turn a pot upside down, tapping it to release the plant without harming the roots.

I hardly knew my own grandmother. The first picture I have of myself as a child was taken in her garden, where I am hoisting a tin watering can big enough to go to sea in. But she died while I was living halfway around the world, so it was her sister, my great-aunt, who taught me to garden. She showed me how to use a taut string to plant in straight rows; how to drop the seeds on top of the earth so I never lost my place, then poke them in with my finger; how to tie tomato stems to stakes in figure-eights. She was over seventy; I was twelve and eager to learn. Since then, I've always had old ladies as friends, old ladies with gardens where I weeded and watered, gardens that supplied their tables, with a few flowers as indulgences: shrubs of roses, window boxes of geraniums.

When I planted gardens of my own, I turned up my nose at roses and geraniums. Old-lady flowers, I called them. But my tastes are changing. I find myself passing up the sugary cookies and reaching for the candied ginger, a shift I notice because this was the sweet my father liked, just as the geraniums I'm potting now were my mother's favourite flower.

These scallop-leaved, scarlet-blooming plants aren't really geraniums, of course. Is anything in the garden ever what it seems? True geraniums—plants that belong to the genus Geranium—are hardy and perennial, unlike these lovelies, which will wilt and blacken at the slightest frost. True geraniums are mostly low-growing and spreading, with the same rounded leaves and flowers that rise above the foliage on thin, high stems. They are sometimes called crane's-bills because of their long, skinny, upthrusting seed pods, which do, in fact, look quite a lot like a waterbird's beak. Oddly, the tender faux-geraniums I overwinter in my guest room take their name from waterbirds, too: pelargoniums, from the Greek *pelargos,* meaning stork. Linnaeus sensibly looked at the rounded, notched leaves and the skinny, upthrusting pods of both these plants and grouped them together as geraniums. Later botanists, however, found distinctions and assigned each its own genus, Geranium and Pelargonium, within the botanical family Geraniaceae.

Despite the family resemblance, there are significant differences. Pelargoniums are annuals in my climate. They have no invasive tendencies at all. And although I love the white-pink-purple colour palette of the true geraniums, if I want a splash of true red in the garden, only a pelargonium will do.

Mostly I see pelargoniums in window boxes, a neat smear of bright-red lipstick on a pale facade, a style perfected in Bavarian schlosses and Swiss chalets. My mother planted hers as a scarlet honour guard lined up in front of the house. I should do that, too, I think. I look around for a square-edged bed, but my soil is exposed in sweeping curves. I could set these pelargoniums in the Hortus Familia, the garden where we grow plants that remind us of our families—a balloon-flower named Blue Moon for my Beloved's father—and where I grow a crazy curly-petalled allium that reminds me of my Sister the Therapist's wildly curling hair.

Instead, I buy a big blue bowl of a planter and pack it tight with the rooted cuttings, hoping that the centre plants will rise tall and those on the outside will spread low to create a Victorian puff of frilly green leaves tufted with blush-red blossoms.

"Small rose. That's what I call it," the Rosarian told me when he gave me this pelargonium the year we moved to The Leaf. The flower is deep red, small and full, like a boutonniere.

How perfect, I think. A rose and a geranium, too. Two lovely old ladies in one.

ALL HAIL THE QUEEN

I FOUND THE SWEET PEAS OF MY DREAMS in a small, stapled-together catalogue from Eternal Seeds, a little company in Quebec that specializes in organic heirloom seeds. *Lathyrus odoratus.*

It's an unfortunate name. Odour is what makes city people wrinkle their noses in the chicken yard; odour is what bursts in a cloud from a rotted potato. Scent is something else altogether.

Gardeners have been bending to sniff this particularly fragrant sweet pea for at least three hundred years. It was discovered growing wild in Sicily by a botanizing monk, Francesco Cupani, who sent the seed to a rare-plant collector in England. Before long, British gardeners were buying it from catalogues, developing such affection for the small fragrant flower that they bow to it as the Queen of the Annuals.

> Odour is what makes city people wrinkle their noses in the chicken yard. Scent is something else altogether.

"It has the sweetest scent on earth!" I tell my gardening friends, pressing on them packets of the 'Painted Lady' seed I collect, offspring from my original planting a decade ago. "And you can't get it in a bottle. This perfume you have to grow."

27

The sweet pea is one of those rare scented flowers that does not produce an essential oil. No chemist can capture it. The fragrance can only be experienced "live." My mother used to say, "Take time to smell the roses," but I can buy rose-scented perfume and soap, even rose-scented tea. She should have said, "Don't forget to smell the sweet peas."

These 'Painted Ladies' make me stop and breathe. They also teach me patience. I remind myself of this as I roll last year's seeds between two sheets of sandpaper (80-grit) to scarify the thick coat, then soak them for a day in tepid water. I have to get to this early, when I'm already busy edging the beds and removing last year's debris, because the seeds can take up to a month to germinate and the plants bloom best if they grow in cool weather. They don't mind a little frost, so in they go as soon as the soil is thawed. An elderly friend of a friend insists that sweet peas be planted on Good Friday. Today is Palm Sunday. Close enough, I tell myself.

I don't really have an ideal place for sweet peas. The Rosarian does. In his garden, the vines twist up the cedar-shingled summer kitchen that anchors the back of their stone cottage, built by the brother of Squire Molton, who built our house. The sweet peas are bathed in sun through the morning, protected by shade all afternoon. The rain barrel is close by, so the roots are kept moist. All summer, scent from the flowers wafts up through the kitchen window, filling the house.

At The Leaf, I planted them first in the south bed by the cedars, where I would smell them as I lolled in the hammock on a summer's afternoon, but the soil was too rich and the foliage overtook the blooms. (And who has time to loll?) Next, I put them in the long perennial border in the West Yard, but they were too far away to perfume the house. Last year, I planted them to climb an iron trellis beside the screened-in Garden Room, where I could read in a heady bath of scent. But it was too hot on that side of the house: the stone captured the day's heat and slow-cooked the stems.

This year, I'll sow them in a planter that I'll set on the east side of the house, on the deck outside my own tacked-on summer kitchen. I'll start early, for the 'Cupani' sweet peas bloom best when they flower before the weather gets too hot. For a trellis, I'll weave branches together with grapevines, and I'll build this support while the plants are still two inches high, just as the books advise. Later, I'll mulch to keep the roots cool, and I'll pick off the blooms as soon as they are spent so the vines will continue to produce. I'll leave the pods to form, heavy with next year's seeds, only at the season's end.

Sweet peas are bred now to be big and colourful, flagrant but no longer fragrant. By comparison, the two-tone, mauve-and-purple blooms of the 'Cupani' sweet pea are small, but even so, bunched together, they make a potent nosegay. It really doesn't matter where I plant them: every morning, I'll snip a few of my 'Painted Ladies,' for a small, round glass bowl with a narrow mouth that I'll carry with me through the day, to my desk, to the sink, to the dinner table; and when the lights go out, I'll set it on the window ledge above our bed, where my Beloved and I will breathe in the fragrance that sweetly scents our dreams.

SPRING TRAINING

MY BODY HAS FORGOTTEN HOW TO SQUAT.

Today we've spent hours cleaning up the gardens, raking off leaves, cutting back the dead stalks, revealing green shards of tulips and narcissi already thrusting through the soil. The Rosarian tells me I start this spring cleaning too soon. "Give the volunteers a chance to sprout," he says. And I would, but the day is warm, with rain in the forecast, followed by driving sun. I am eager to cast off my winter clothes and bask. Why keep my seedlings under wraps?

But my gluteal muscles are complaining. How soon they forget the particular postures of the garden. That rump-in-the-air bend from the waist as I dig out the quack grass and dandelions. And the more modest and more demanding squat, knees splayed, back straight, bum inches from the ground, the only position from which I can comb both hands through the soil, demolishing the forest of forget-me-nots.

I intended to be sensible. "Just an hour," I said to my Beloved as we headed outside into the newly balmy breeze. "We'll start slowly."

And we did. By the time we located our spring gardening gear and brought the tools from the shed to the cupboard where they live through the gardening season, it was already mid-afternoon, the sun a faint memory behind a bank of heavy cloud.

"Three beds," I said, taking care not to look at the other twenty-three.

"Three," he agreed. "Then we go in for cookies and tea."

"Tea," I corrected, feeling querulous. "There aren't any cookies."

He trimmed back the Preston lilac while I cleaned the Rockery that spreads out over its roots. This is the wrong time of year to prune a lilac, I know, but the shrub's skeleton is fully visible now, its limbs deceptively erect. Burdened with flowers, they'll bend almost to the ground, shading out the other plants. I directed his cuts, trying to imagine the leafed-out limbs, pruning them high in what we hoped would be a pleasing vase shape.

The work went swiftly. While he carted off the trimmings, I scraped last year's debris from the Rockery, uncovering a lone purple crocus, a nest of pasqueflower like fuzzy catkins curled in on themselves, waiting to stretch up into the sun that is bound to return. I pulled out handfuls of the vigorous yellow coreopsis my Garden Guru gave me, restricting them again to their designated spot, then broke off the withered branches of the delicate 'Moonbeam' coreopsis and crushed their seed heads over the soil to encourage them to propagate. Managing these two coreopses is like parenting mismatched twins, gently encouraging the introvert, reining in the extrovert, loving them both.

We straightened the edges of the garden as we went. It's easier at this time of year, when the grass is tender and the soil moist. We worked from the sidelines, leaning into the beds, the ground still too damp to support our weight. My Beloved ran wheelbarrow after wheelbarrow of gleanings off to the compost, still frozen in its wooded corner of the yard. I thrust in the spade, slicing through a mound of clematis, heaving the offcut like a shotput into its new hole. I tossed the rake across the herb bed to my Beloved. He caught it mid-air and lobbed the Japanese knife back to me, sinking it expertly into the ground at my feet.

We should train for these spring games. The caber toss, stone put, and sheaf toss have nothing on what we've been getting up to on this March afternoon. We came to the garden with our fat and flabby winter bodies, muscles unwarmed, blood sluggish, lungs weak after months of sipping shallowly at indoor air. Then, mindless of the ripping and tearing going on under the skin, we leaned and crouched, tossed and flung, pulled and pushed and heaved.

"Time for tea," my Beloved says sternly as the light begins to fail. He holds his rake before him like a crossing guard with a stop sign.

"But we've only done two beds!"

I scan the yard. We were travelling last fall; nothing has been done in the gardens since September. No weeding. No mulching. The beds stretch to the woods, a dark jumble broken only by bold green spires of dandelion.

He grasps me by the shoulders and turns me away from the chaos to face the Kitchen Garden, a smooth sea of soil divided into four neat squares by wood-chip paths. Welsh onions bunch on either side of what we think of as the entrance. Oregano, thyme, sage, and rue, neatly trimmed, flank the west arm. Pots of lemon balm hold the centre of each of the south beds. The bee balm is rising. Even the clump of catnip looks plump, despite the glaze-eyed rolling of the cats. The rose bushes that guard the exit are clipped evenly to their first swelling buds.

Gently my Beloved points me to the Rockery, where the lilac is a vase ready for blooms, each plant at its feet sweetly framed in dark earth.

"Look," he says. "We've *already* done two beds."

But I'm not looking at the gardens. I'm thinking ahead to tomorrow, grateful for the rain that is promised. Rain that will settle the plants we've dislodged from the weeds; rain that will keep us indoors for the day, where I will demand nothing more of my muscles than to climb the stairs to my office and lean over my desk.

"These two gardens look great," I say. "I'm a little sore. How about you?"

He puts his arm around my shoulders and guides me down the stone steps to the house. "Like I said, time for tea. There may not be cookies, but I stopped at the bakery." He winks. "We have hot cross buns."

PASQUEFLOWER

When I was growing up, Easter meant new clothes. A spring coat. Maybe a hat. Shoes, for sure. And the year I turned thirteen, white gloves with pearl beads for buttons. Women still wore gloves in public then, to church and cocktail parties; my silky white hands were a sign that I was a grown-up now, too.

There was no Easter parade in my village, nothing like the throngs of men and women, a million—sometimes more—who strolled down New York's Fifth Avenue displaying their spring finery. But there was the walk to church in our new, sweetly coloured outfits, such a relief from the dull, dark hues of winter. April in those days was inevitably cold and windy, so the fabrics, though bright, were fuzzy and warm: boiled wool in shades of lavender, a fluff of lambskin at the collar.

I think of those spring coats when I see the pasqueflower rising in late March in the Rockery. The buds are soft and grey as lambswool, curled into themselves. The leaves, when they come, will be greyish green and lacy, covered with the same silvery, silky hairs that soften the stems and buds, but it is the flowers that show first, pale purple cups with bright yellow stamens at the centre, the colours we painted our Easter eggs. Sepals in darker shades of purple frame the flowers, and on their undersides, a second set of leaves shields the buds against late frost.

It is a cozy arrangement. The sepals raise the temperature within the flowers by several degrees, enticing small insects to crawl inside to warm up. Bees, starved in these early days of spring, cavort on the stamens.

No one I know grows pasqueflowers. Perhaps they are too wild for cultivated gardens. Or too confusing. Two different members of the buttercup family claim the name—the Old World pasqueflower, the purple *Anemone pulsatilla* that reminds me of spring coats, and the American pasqueflower, *Anemone patens,* a bluish wildflower of the tall-grass prairie that was given the name because it resembled its European cousin. *Anemone patens,* which is also known as Prairie crocus, gosling flower, and meadow anemone, is the official flower of both South Dakota and Manitoba, though the Americans call it *Pulsatilla hirsutissima,* a name the official referees have declared illegitimate. Local Natives ignore the Linnaeans and call it *hosi cekpa*—child's navel—maybe because it begins as something that looks like a belly button, or maybe because they once used the plant, which is mildly poisonous, to induce abortions.

It is the Old World pasqueflower that I love. *Anemone pulsatilla,* or as some refer to it, *Pulsatilla vulgaris.* For years, I've wanted it in my garden. I tried growing it from seed, but germination was poor. I tried planting seedlings bought from nurseries. They'd thrive in the first season, never to return. I set them in the sun, in the best soil I could find. Still they failed to last the winter, which seemed odd to me, since the plant is supposed to be hardy to zone 4, easily within my range. Finally, last year, I put a pasqueflower in the Rockery, in my poorest, stoniest ground, the closest I could come to the limestone pastures of central and northern Europe and the parts of Russia it calls home.

And when I cleaned the Rockery this week, there it was, a mound of grey nubbins, soft as down. New growth must have started the minute the snow melted. In a week, just in time for Easter, there will be blooms, pale lavender, the centres white as the gloves I used to wear to church.

There is something endearing about the first blooms of spring—the snowdrops and crocuses and woodland ephemerals. They don't have to be pretty to win our hearts; like children, they simply have to show their soft faces. The pasqueflower, though, lingers on into June. Long after the snowdrops have dropped and the crocus petals have withered to translucent tissue, the seed heads of the pasqueflower are waving in the breeze, a nimbus of silvery threads that reminds me a little of my own unruly white curls, and I think, Maybe it's time for a new lavender coat.

GARDEN DWARFS

EVER SINCE THE MOVIE *AMÉLIE* I have wanted a garden gnome. A garden dwarf. A little person who lives in the garden. The antiques-store owner had no idea what I was talking about, though I used my hands and even drew what I thought was a pretty good likeness. I was on a reading tour in Munich, with a day to myself. Where better to look for garden sprites?

What I had in mind was something quaint, made of wood, painted in rich dark hues, the expression happy and wise, maybe a little sardonic. A droll troll.

The inspiration came from a friend who told me that when he was young, he lived next door to an old man with a large and magical garden. A wooden gnome inhabited the low hollow of a tree, and as a child, my friend would find there notes addressed to him, little homilies and encouragements—*Be happy today. The sun is shining!*

I had the idea that I would conjure for my Grand Girls the same sort of magic at The Leaf. I imagined the little fellow sitting among the snowdrops or with miniature daffodils at his feet, adrift in a field of blue scilla. Set beside a garden gnome, these small flowers would seem monumental, as grand as the pleasure they give me.

Little bulbs, that's what the American writer Elizabeth Lawrence called these miniature blooms. I wish I'd known Elizabeth, who would have been in her seventies when I started my first garden. I've always been drawn to older women, and she was a diminutive Southerner, direct and gracious, the way I like them. She wrote an entire book on the subject—*The Little Bulbs: A Tale of Two Gardens*—discussing everything from the earliest snowdrops, squills, and daffodils to the fall-hardy cyclamens and crocuses, including in her treatise the wood sorrels, the little irises—ixia, tritonia, freesia—and all the members of the lily family—mariposa tulips, glory-of-the-snow, fritillaria, hyacinth, trillium, wild tulips, clusiana tulips, and more.

There's something quaint and precious about a daffodil the size of your thumb, an iris that fits neatly in a button-hole. Child-size flowers. You have to get down on your knees to appreciate a squill, and that changes everything. Pebbles become boulders, earwigs grow big as lobsters. A chipmunk wouldn't carry a snowdrop as a parasol; he'd put it in an acorn-vase on his thread-spool table.

The garden catalogues call them dwarfs. They have the same effect on me as the Grand Girls—they make me want to play.

In *The Secret Garden*, when Mary first pushes open the gate and sees the sharp, pale sprouts in the flower beds, she whispers, "They are tiny growing things, and they might be crocuses or snowdrops or daffodils." Later she asks the housemaid, "Do bulbs live a long time? Would they live years and years if no one helped them?" And the maid replies, "They're things as helps themselves. That's why poor folk can afford to have 'em."

When I lived in the city, I used to visit a dilapidated house every spring. The front yard was small and dark, deeply shaded by maples left unpruned for years. If I were a child, I'd call it a haunted house. It had once been one of those Victorian grandes dames and was painted a faded, though faintly cheerful, yellow that in April fairly glowed behind the bluest lawn you could ever imagine. Pools of scilla dot the city, but nothing like that old dooryard, where the small blue flowers had reseeded unperturbed for generations until the spring grass was a dazzling great blue sea. Things as helps themselves, indeed.

Squills is what the Brits call this little star-shaped bell-flower, a word that sounds pointy and scratchy, though it has no meaning beyond that pretty little plant. I prefer its Latin name, *Scilla*, which makes me think of Scylla and Charybdis, the monsters that haunted either side of the narrow, rocky strait that Odysseus had to sail through on his endless voyage home. Steer clear of Charybdis and he'd pass too close to Scylla, a hideous sea creature with twelve tentacle legs and six long necks, each with its own grisly head baring three rows of gnashing teeth.

I tell the Grand Girls the story of Odysseus's voyage, which they find funny because our cat is named Odysseus. We try to imagine why these delicate blue flowers are named for monsters.

"Because the stems are long—like necks," says the younger one.

"Because they're the same colour as the sea," says the other, whose favourite colour this year is blue.

"They come from Persia," I say. "A Persian princess once married a king who lived far away, and she missed the gardens of her homeland so much that her husband, King Nebuchadnezzar, built her the Hanging Gardens of Babylon, one of the Seven Wonders of the World."

I'm a little shaky on my history, but they don't care. They love the story, and scurry off to draw hanging gardens of their own.

I have been trying for years to grow a magic carpet of *Scilla siberica*. Native to the steppes of northern Persia and southern Russia, it does well in this part of the world, too, where it gets just enough cold to make the stems rise, lifting the flowers high above the grass, where they fade just before the first mow. I bought six bulbs six years ago and I must have six hundred now, the thin leaves arrowing up in beds where the wind sows the seed, though I help, too, casting the heads in handfuls through the gardens and over the grass.

I may regret this. When we first moved to The Leaf, I found a small bulb growing in the lawn. It pushed up leaves but no flowers, so I moved it to a sunny bed, thinking perhaps it was scilla brought on the breeze from someone else's yard. The bulb flourished and multiplied, blooming white in mid-spring. Within a few years, the soil was clogged with hundreds of bulbs the size of pearl onions. I moved them into the shade, hoping this might slow them down, and it did, but only a little. Now I find them everywhere, among the asparagus, in the Pine Garden, under the rhodo that never blooms, even sprouting from the edges of the compost.

Most small bulbs have delusions of grandeur. They multiply like crazy. Still, I keep planting them. Neighbours drop off clumps of snowdrops. At a yard sale I pluck a plastic bag filled with freshly dug blue muscari placed under a sign that reads FREE TO A GOOD HOME. I buy nameless dwarf daffodils at the grocery store through the winter, enjoying them for a few weeks, then letting the foliage die down until spring, when I enlist the Grand Girls to help me put them in the ground.

"Around the birdhouse," one says, meaning the small, copper-roofed pergola that sits in the Pine Garden as if waiting for the Borrowers to drop by for an outdoor picnic.

"By the playhouse!" says the other. "Under the climbing trees, by the princess tent!"

The only dwarf species I buy at nurseries are tulips. I've always disliked tulips—not the flowers themselves, but the irritating tendency of these big, hybridized bulbs to weaken year by year until the bloom is a frail imitation of itself. Instead, I have become an aficionado of species tulips—botanical tulips—the small, wild parents of those long-stemmed, persnickety hybrids, which I have come to treat as annuals. It was species tulips that got the Dutch bulb industry going in the sixteenth century, and they are still out there, growing in faraway mountain ranges, hidden gorges, and remote meadows: low-growing, small-flowered tulips of intense colour in a dizzying array of petal shapes.

One of my favourite garden dwarfs is just sending up its leaves now. I don't remember where I got these miniature irises; someone gave them to me when I lived in the city, and I've been moving them bed to bed ever since, searching for a spot where we're both happy.

"Come, smell these," I say to the Grand Girls, and the three of us crouch down on all fours, snuffling the air like strange beasts.

"It's chocolate!" says the younger one.

"This one is vanilla!" says the other.

"Can we eat them?" they chime.

"No, these are just for smelling. They'll make you sick if you take a bite."

"Like Snow White and the poison apple?"

"Something like that," I say, picking bouquets for my mother's tiny cut-glass vases, one for each of the girls.

My greed for garden smalls is insatiable. I love the way they drift into odd corners, spread themselves under the viburnum, the lilac, the snowball spirea. Those few dozen bulbs of *Muscari armeniacum* have enlarged from puddles to pools of blue in the Rockery. I can hardly wait for the leopard lily, a furry, spotted bloom that peeks above the summer flowers. I long for a dwarf calla lily and a miniature dahlia, just a foot high and across, covered with thumb-size rosy blooms. And the lovely egret flower, *Habenaria radiata,* with gleaming white flowers that look like birds in flight.

MAKE A WISH! the Grand Girls write in mismatched letters with red and yellow crayons on a piece of paper I find tucked under my garden gnome. I discovered him at last on the back shelf of a German grocery store, right beside the rubber rings for the glass-topped canning jars that line our pantry cupboard, a king's ransom of ruby beets and heart-of-gold squash. My gnome stands less than a foot tall and is dressed like a fairy-tale woodcutter, with a pine cone for a cap. He is more of a forest dwarf than a garden troll, I suppose, but he has a kind face and a sprightly stance, as if unaware that his plaster flesh will slowly dissolve with the seasons. The Grand Girls have built a house for him, a strange weaving of sumac whips on the broad stump of an ancient elm. They leave messages for me there, drawings of stick-rayed suns and smiling birds and flowers that rise above the grass—which they colour a scintillating scilla-blue.

ONE CROCUS

WHEN I WAS THREE, MY MOTHER dressed me in a bunny costume and sewed me an orange-felt carrot with green-ribbon leaves that I held on to for dear life.

I know this because there is a photograph. In it, I am grinning as if a smile is my only hope. Pasted into the album beside the picture is the musical score for "The Crocus Song." I don't remember dressing up or performing the piece for Baby Band, though I can still sing all the words of every verse:

"The crocus flower peeped out, to see if it was spring . . ."

I have planted hundreds of crocuses—croci?—in my gardening life. Maybe thousands. Pale cups designed to erupt through the lawn as the first grass is greening, and if not there, then in the garden. The sort of thing you see in British horticultural magazines, a wood of stately trees with low, flowering limbs, and on the ground, a crazy-quilt of croci.

The crocus, once a member of the lily family and now an iris, comes in two types, according to when it blooms, spring or fall. The fall-blooming crocus sends up in the spring leaves that die down in June, disappearing so completely you forget you planted anything in that place until, in September, an unlikely mauve bloom thrusts up its head, a guest arriving late for the party.

The spring-blooming crocus behaves more predictably. It sets its bloom first, then sends up leaves to gather nutrients for the bulb so it can bloom again the following spring. If the foliage is left to develop and die down naturally, the corms at the base of each bulb increase year to year, until there is such a mass of blooms that the whole lot has to be lifted, separated, and replanted to give each some breathing room. Or so I've heard.

Although I like the idea of colourful crocuses dotting a sweep of spring-green lawn, I can't bear the thought of lopping off the leaves before they have a chance to feed the bulb. This would inevitably happen, because my Beloved's favourite part of the garden is his well-mowed lawn. Instead, I plant my crocuses out of harm's way, in the perennial beds, judiciously poked among the tulips, the daffodils, and the hyacinths. The best effect, the books say, is obtained when crocuses are massed in a medley of yellow, lavender, blue, and white, and so I open the packages and mix the colours, a painter with her palette. I squint at the soil and imagine a Monet moment, the rainbow blooms blending into a hue that could only be called *spring*.

I plant and I plant.

Only one bulb sends up a flower. It opens on the last day of March, a translucent, pearly white.

I am frugal but not cheap, an important distinction.

"One crocus is no crocus," my Beloved sniffs.

I plant more. I am frugal but not cheap, an important distinction. I have long since learned the most basic of the gardener's lessons: you must sow before you can reap. The cost of the bulbs is an investment, I tell myself. Before long, I will be awash in crocuses. I'll give basketsful to my friends, force the bulbs in pots on every winter windowsill. And so I buy more. Many more. In the fall, I shove them by the handful into soil already too cold for comfort, warmed by a vision of the Easter-basket colours that will announce the new growing season.

Come March, one blooms. It is white.

Squirrels may be the culprits, my Garden Guru suggests. We do have squirrels, but only the small, ruddy pine squirrels, and they show little interest in digging. They prefer to harvest cones from the conifers. All summer and deep into the fall, they sit at the tip of one pine branch or another, high above the terrace, stripping off scales like artichoke points that they nibble before raining the gnawed bits on our heads as we eat our lunch.

I prefer to blame the chipmunks that live behind the stone wall in a maze of tunnels I break into now and then when tussling with long-rooted dandelions. We never see the chipmunks in the winter. Around the twenty-first of March, a sentinel will show itself, poking up through the creeping phlox, swivelling its head like Wiarton Willy on a sunny Groundhog Day. Soon, a steady brigade of racing-striped rodents are pell-melling back and forth to the feeders, where the profligate blue jays have left morsels of nutmeat among the scattered shells. From there, they go on to the lilies, stripping the blooms just as they are about to open. They pick my strawberries and crouch on the stone wall, holding the fruit to their mouths with little paw-hands, mocking me, waiting until I look, then savouring my treat in full view. I'm more than willing to hold them responsible for the missing crocuses, too.

Mostly I plant the bulbs near the house, on the slope under the red pines, the slope that ends at the stone wall overhung with *Phlox subulata* and mother-of-thyme. I imagine the chipmunks in midwinter, tugging on the bulbs from below, sinking their tiny white teeth into that succulent flesh, firm and juicy as roast turkey.

We'll leave her one, shall we? the wily elder says.

Okay, but just one, the others agree.

I have read too much Beatrix Potter, seen too much Disney: the chipmunks nod their heads under cunning little caps made from acorns. Knitted scarves of milkweed floss wind around their necks.

Once again this year, I watch the solitary cup open—white, striped with mauve—seeing in that tender bloom the promise of my endless field of spring.

"One crocus is no crocus," my Beloved says again.

Where the garden is concerned, I seem incapable of despair.

"One crocus is every crocus," I reply.

MOVING DAY

I COULD SEE THEM OUT OF THE CORNER of my eye as I dug up the quack grass and the dandelions and cleared off the last of last year's weeds. Columbine growing up through the golden juniper. Bleeding heart rising in the sliver of space between the downspout and the house. Where the dry stems of last year's astilbe clatter, a bare spot, perfect for primrose. And that empty corner under the pines, how it begged for a burst of 'Sum and Substance' hosta to draw some light into the shadows.

"Time to get moving," I said, pulling on my gloves.

"Like Montreal in spring," my Beloved mused. "On May first, everyone moved, whether they needed to or not. It was a tradition."

I've seen the photographs: wagons and trucks piled high with boxes and dressers, mattresses roped on top, streets clogged with conveyances, people riding their possessions like floats in a parade. There was a localized version in the city where we used to live, too. On May first, the students would pack up and move on, going to jobs in other places or home to their childhood rooms. The streets of the student ghetto would be lined with U-Hauls, lawns heaped with sprung sofas and collapsing desks, stuff too worn and wrecked and too full of memories to carry into their new lives.

"I can't wait until May first," I mumbled, rummaging in the outdoor cupboard for my transplanting spade.

But I had to wait, at least for a while. One thrust of the spade and it was obvious the soil was too dry. I dug down eight inches: hardly any dampness at all. No rain in the forecast, either, and the mercury rising. It was midsummer in April. Anything transplanted now would wilt and falter, maybe die.

All week, I watched the sky. I made a list that grew with every day. *Divide primrose—move to slope among the pines. Move purple columbine behind meadow rue.* Its blooms would fade just as the rue rose to hide the last leaves, or that's what I imagined. *Split and move chartreuse hosta. Move burning bush.* Last year, one side failed to leaf out, the cedar had crowded in so close. *Lift and split daffs—move to the Shrubbery.* Move. Move. Move.

This dance appeals to me. In the winter, when I feel downhearted, I shove the couch and chairs around, pack up the wooden duck decoys and ceramic inkwells and haul out the art nouveau balancing man and sky-blue enamel bowl, shift the paintings on the wall. "A change is as good as a rest," my mother used to say. After rearranging the furniture, I carry on refreshed.

I eye the sky. I need rain, and not just any rain. A transplant day: overcast, with a solid tea cozy of cloud. Warm enough to make the work pleasant, but not too hot. Dry in the morning—a few hours is enough to get the plants in the ground—then a light, misty precipitation, one of those gentle rains scarcely felt on the skin, a rain that goes on for hours, into the night and through the next day, too.

The temperature lowers a little, but the sky remains dry-eyed.

I console myself with volunteers. I don't invite these into the garden. They come of their own accord, surfing in on the breeze or riding like little Jonahs in the bellies of birds, sprouting where they land, mindless of my careful planning. The white astilbe that took root behind the Korean lilac: no one can see it there. And the orange asclepias that blooms among the red peonies: it will be much happier with its fellows at the foot of the Ohio buckeye—and so will I.

I scoop them up with my Japanese gardening knife as I come upon them and drop them into pots I keep stacked by the outdoor tap, the one with a brass bluebird for a handle. As April moves through its first wayward days, the crowd of pots waiting by the tap grows, a homeless population of migrants. I've taken little care in potting them up—a scoop of soil, the plant pressed in, a second dash of earth to cover the roots. Sometimes less. As I cut back the stems of 'Annabelle' hydrangea, I pull up a root that has thrust too far afield and toss it into a big black pot, covering it with last year's leaves, all I have at hand at the time.

"Hydrangea can stand it; they're tough," I say brusquely when my Beloved asks what I'm doing watering a pot of dead leaves.

I shouldn't be so cavalier. When I was a child, I lived in Brazil. My family returned to Canada when I was twelve, to a small town at the heart of the country. Everything about the place seemed wrong to me. I never did fit in, not in my own mind, though to outward appearances, I suppose, I thrived. But there has always been a longing in me, a place where I felt something missing, the way an amputee feels a limb that is lost. A root torn up, broken off.

It is raining at last. Not as warm as I'd like it, and not as gentle. It must have rained all night: the soil is already soaked. It is too wet to walk on, but I can work at the edges of the gardens. As I pull on my yellow slicker, I go through the list. First the perennials. I'll prepare the new ground, working it with the Garden Claw until it is light, then cut through the root cleanly or dig carefully around it, lifting it so soil still clings to the smallest root hairs. I'll note the direction the leaves face and position the plant so the same surfaces bend to the sun.

Move quickly, I tell myself. Disturb as little as you can.

After the perennials, the volunteers. I cram the pots into the wheelbarrow, my red-tin immigrant ship, and make the rounds of the bare spots, setting plants in, digging up others I notice along the way.

My Beloved watches, bemused. "Didn't you move that last year, out of the same place where you're planting it now?"

I look at the clump of white penstemons. Could he be right? Am I just shifting the furniture around, rearranging the plants because it's spring and I crave a change?

As I bend to set the clump in place, wet soil falls from the root, tearing away the fresh white growth. I settle the plant as best I can, pat the soil firmly around the stems. My hands are cold. The clouds have turned a deep bruised blue. The air smells of snow. I always thought that I was working toward a vision for these gardens, that this moving about would one day end. I don't like to think I'm disrupting them, maybe causing them harm, on a whim.

I stand and survey my morning's work. The crowd is gone from around the tap. The bare spots are green. If it doesn't snow, it will surely rain. Either way, there will be moisture. I tell myself the plants will be fine; they'll survive. I can't help it: I look around and feel happy with the new order of things.

"Maybe it *was* here last year," I say, smiling at my Beloved, who has a sweet smirk on his face. "But it looks good, don't you think?"

GRASSPLOT

MEN LIKE GRASS; WOMEN LIKE GARDENS. So my Garden Guru says.

"How can you like grass so much?" I ask my Beloved. "Grass has no flowers. It has no scent, unless it's cut and dead. It respects no boundaries."

"Precisely," he replies.

When our house at The Leaf was built in 1824, only Old Country dukes and earls had lawns. Settlers had dooryards planted with peas and cabbages, maybe a swath of greenery beyond the stoop that was scythed now and then or nibbled by the family sheep to keep the wilderness from creeping too close. The scythed area was never very big. What with felling trees and planting crops and managing the livestock, who had the time?

When we arrived at The Leaf, the lawn stretched grandly in all directions, covering two acres at least.

"The people who lived here must have been in love with their lawn mower," I said.

"Or with Capability Brown," added my Beloved.

Capability Brown was a landscape architect in Britain in the 1700s, a man some call England's greatest gardener. He was only seventeen when the word *lawn* was coined for a mowed grassy plot, so I can't attribute it to him, but he, more than anyone, is responsible for our North American obsession with green-plush yards. It was Brown who gave us a taste for what he called "the pleasing prospect," that sweep of undulating grass that leads the eye to a specimen tree, an arrangement of shrubs, a fountain, the front door of a house.

A famous portrait of Brown shows him cocking his head at the viewer, a glint in his eye and a barely suppressed grin on his lips, as if he knows the havoc he is about to wreak, as if he can already hear the arguments between countless husbands and wives. *More gardens! No—more grass!*

Still, I find it hard not to like a guy who describes a landscape in terms of grammar.

"Now there," he'd say, pointing a finger, "I make a comma, and there, where a more decided turn is proper, I make a colon; at another part, where an interruption is desirable to break the view, a parenthesis; now a full stop, and then I begin another subject."

I share Brown's affection for punctuation, but not for the gardenless form of landscape design that is reflected in every contemporary suburban plot. Grass stretches around these houses like a mat around a painting, drawing attention to the prized human construction at the centre. In Brown's horticultural vision, borders of flowers, trees, and shrubbery served the same function as plaster roses on a fancy picture frame.

At The Leaf, the mat inside the frame was enormous: mowing the grass was a ten-hour trial.

"It's too much!" we both exclaimed, exhausted.

We stopped mowing the meadow and cut paths through it instead. Every year I dug up sod to plant shrubs and flowers and, yes, ornamental grasses. In the third year, the Garden Guru introduced me to lasagna garden-making—twelve sheets of newspaper, topped with a layer of mulch. The following year— *presto*—soil, which meant I could eat up the lawn even faster.

Within five years, we'd cut the mowing time in half. Then we bought a riding lawn tractor with a wide cutting blade, and "cutting the grass" shrank to an ordeal of little more than three hours. We mow the lawn ten times a year. (I've kept track. Wet year, dry year; early spring, late fall: it's always ten mows.) That means thirty hours on a noisy, spewing machine whose only saving grace is that it blows the clippings into bags so I can use them as mulch.

If a garden teaches anything, it's balance. Give and take. My Beloved loves the lawn. I hate the mower but crave the clippings. And so we've struck a bargain. I concede what's left of the lawn. In return, he promises an endless supply of mulch to smother the perpetual weeds.

I honour the agreement. No more gardens. I leave the sod where it lies. I resist a sudden, inexplicable craving for a rose garden to border the Croquet Lawn. A deal is a deal.

At least, it is for me. The following spring, the grass launches an assault. From a thousand landholds, it sends runners into the garden beds, staking new ground. The quack grass is the worst. In winter, it dies back to its rhizomes, underground stems that look like long, pale strands of alternating ebony and ivory. Every link, even a piece of creeping root stalk as small as one of those jewellery pop-beads, can make a new plant.

With grim satisfaction, I set out in April to redraw my line in the grass, pulling on the stripey roots, feeling them slip through the soft, moist earth. I capture rhizomes as long as my arm, skinny snakes that I stuff into black plastic garbage bags, where the pale sun cooks them to slime.

"There's a machine for that," my Beloved points out.

I know. I've considered the lawn edgers at the rent-all where we get the power pruners to bring the cedars down to size. I'm not a Luddite, but it occurs to me that if we'd stuck to scythes, lawns might never have got the upper hand. If only Edwin Budding hadn't visited that cotton factory in 1827. If only he hadn't seen that nap-trimming machine. If only he hadn't thought, A roller like that fitted with blades could cut the grass! (A woman looking at the same machine would have invented something truly useful, like a sweater-shaver.)

From the beginning, Budding's mowing machine was advertised for its personal appeal to a man as much as for its efficiency in lawn maintenance: "Country gentlemen may find in using my machine themselves an amusing, useful and healthy exercise." The first mowers were pulled by mules; I wouldn't mind that so much. I once suggested to my Beloved that we keep sheep to trim the grass.

"They'd fertilize as they mow," I said.

"And who is going to keep the coyotes at bay?"

My Beloved likes machines. I prefer old-fashioned hand tools, like my edger, a Cornish pasty of stainless steel on a handle that rises to waist height with a crosspiece at the top. The blade is slightly flattened on its straight edge to take a sturdy boot.

There is a satisfying rhythm to edging. With one boot on the blade, I rock my way along the border of each bed, slicing down through the advancing grass, holding it to the line. I have a good eye. I never use a string or mark the end-point with a stake. I spot ahead where I want to go and rock my way toward it. After the line is cut, I go back and lift out the sod with a fork, getting down on my hands and knees to trace the quack grass roots as far as they go.

Once the line is cut, I gaze back at what I've done. Often there is a swell to the smooth curve that needs correcting. I recut the line, edging an inch or two deeper into the grass.

"That's where the line was," I say out loud, staking a claim of my own, "before you moved it."

When I cut around the trees, I am generous. For the sake of the mower, I say, so the curve can be cut in a single sweep. Occasionally, the tree-bed touches the edge of a border, and then what choice do I have? I redraw the line again, moving it forward to embrace the Ohio buckeye, the Kentucky coffeetree. I plot my moves with care, proceed with subterfuge. Every spring, it seems the grass has won the battle. But the war isn't over yet.

"The gardens are getting bigger," my Beloved muses of an evening, looking out across the smooth green cloth of his freshly mowed grass.

"Are they?" I reply. "I hadn't noticed."

HEARTSEASE

WHEN MY YOUNGER SON MARRIED, I made the wedding cake, two layers of sherry-soaked sponge spread with a white chocolate ganache, embossed with a white chocolate lattice and foliage that I made by brushing strawberry leaves with three thin layers of liquid chocolate, then peeling off the plant material. On top of the cake, a silvery cup offered up a bouquet of wild, sweet-scented violets.

Every year when the violets bloom I think of him and his wife, that lovely morning in the backyard, and the pleasure of walking the hedgerows, bending to pick flowers for their cake, bundling them in damp tissue for the long ride to their city.

I think of my mother, too, because violets are cousins to the violas and pansies she loved. Flowers with faces, she called them. The summer Sundays of my childhood are dotted with tiny bouquets of mauve face-flowers set in one small vase or another on our breakfast table; vases handed down from her mother to her and now to me; vases that I fill with violets in the first weeks of May.

Violets, violas, pansies—they're all members of the viola family. *Viola sororia,* the common blue violet that is native to eastern North America, is the one that gives our lawn its lovely sky-blue cast in early spring. It has no scent, unlike *Viola odorata*, the violet-coloured bloom introduced from Europe and Asia and known variously as sweet violet, English violet, common violet, and garden violet. *Viola tricolor* is the one my mother called heartsease, a much more appealing name than field pansy. Heartsease is native to Europe and her Scottish homeland, but it has naturalized in North America, as has the scented violet, which crops up in a different place in the garden each year. Commercial garden pansies are a complex hybrid of several species of viola, including *Viola tricolor* and *Viola lutea,* the mountain pansy.

All violas have five-petalled flowers: four that sweep upward in pairs and one that is larger and lobed and hangs down like a cleft chin. This, together with faint markings that give the upper petals the look of eyes, makes it easy to see why Lewis Carroll included them in Alice's Garden of Living Flowers.

At The Leaf, sweet violets grow wild, colonizing every bit of bare earth. I have no idea how many species grow in my gardens. I've moved them in from the woods, from the lawns, from under the hedges—yellow, blue, pale mauve and deeper purple, striated, white. Most gardening websites classify them as weeds and offer tips for killing them off. "Remember," warns one, "eradication is the watchword with wild violets."

Poor violets. Compared to garlic mustard, they are goddesses to me: fecund, easygoing, they generously spread themselves over the ground, offering endless spring bouquets.

Violets reproduce by those creeping underground stems called rhizomes, and sometimes by runners above ground, some developing in clumps, some in ever-expanding mats. Not for nothing is the native violet called *sororia*; the plants stick together like sisters. But they're easy to keep in check. Early in the season, I rip out the front-runners, confining the mat to where it belongs. In the Hortus Familia, I've trained the violets to a wide border that leads sweetly from the driveway into the yard.

Like most truly successful plants, violets don't stake their lives on just one means of reproduction. They're bisexual, which means that as well as the rhizomes and runners, they produce two kinds of flowers, both of which develop seed. The gorgeous spring flowers that bloom in May—the chasmogamous flowers—develop seed pods that dry and burst, scattering seed up to four feet away. In summer, if you bend close, you'll notice another clutch of smaller blooms at the base of the plant. These are the cleistogamous flowers—tiny, closed, self-pollinating flowers that incubate extremely abundant, fertile seed, thus ensuring the plant's survival in the event the first spring seeding fails.

Some consider violets the Trojan horse of herbaceous plants—let one into your garden, and you'll soon be battling a thousand—but I've found a heavy mulch usually smothers the seed and the runners are easy to control. In return for scant labour, I reap not only Sunday breakfast bouquets for the table but sweet flowers and heart-shaped leaves for the salad bowl, too.

As much as I like its cousins, I've never planted commercial pansies. Their big painted faces seem too sad to me. When I was young, I stayed for a time with my great-aunt and her brother. My great-uncle wore wide suspenders and was hard of hearing from the planing mill he'd worked in as a young man. Before that, his job was painting decorative stripes on the sides of horse-drawn buggies. By the time I knew him, he rarely remembered who I was. Every morning that summer, he'd fill a saucer with milk and shuffle out to the bank of pansies that waved beside the back door.

"They're so thirsty," he'd say in consternation, when I asked why. "Just look at their little faces."

ONE YEAR'S SEEDING

THE PERFECT WEED

YOU CAN HARDLY OPEN A GARDENING BOOK without coming across this pious injunction: *A weed is just a plant growing in the wrong place.* As if that is supposed to make us love quack grass, dandelions, thistles, nettles, the poisonous cow parsley. Weeding, these books imply, is just a case of anywhere-but-in-my-backyard.

But aren't there weeds that are vile by their nature, weeds that should be ripped out no matter where they grow?

Garlic mustard, for instance. Surely *Alliaria petiolata* deserves a place on anybody's list of the ten least wanted weeds. This clumpy bouquet of blackish green, serrated leaves sends down a taproot with a devilish S-curve at the top that breaks off with the slightest tug, leaving the full root in the ground to effect its resurrection. The first-year rosette is quite pretty, really, with those crinkly leaves. The plant stays fresh long after the rest of the garden has withered, and it is the first to green in the spring, thriving in the cool damp of the April garden and flowering while there's still lots of open ground to accept its copious buckshot seed. Sixty-two thousand per square metre. (What poor nameless soul counted?) Uprooting the plant solves nothing: the flowers continue to develop, producing offspring long after the parent is dead. As if all this weren't enough to ensure survival, the seed remains viable in the soil for years.

And what does this super-weed live for? It lives to kill. You'll find it cozying up to the stems of the lovely spring ephemerals, the woodland flowers that open in succession like pastel fireworks through April, May, and June. In the garden, it hugs the climbing clematis, leans up against the stalwart lupin, insinuates itself among the rhizomes of the Siberian iris. Once the ground is disturbed and the seed falls, seedlings rise in a choking carpet, though strangulation is not the worst of it. I suspect there is something female in its spirit, because it murders with poison, a woman's weapon. The roots leach a vile chemical that destroys the mycorrhizal fungi that ensure healthy growth in forest flowers and trees. And the poison is persistent. Maple saplings planted in ground that was garlic mustard–free for two years grew into stunted, twisted versions of themselves.

It's our fault. We brought it here, my Beloved and I. Not intentionally, the way the first Europeans did, planting it for the spicy garlic flavour it lent to soups and for its tonic properties. They called it sauce-alone. Jack-in-the-bush. Poor-man's-mustard.

But we are villains, nonetheless. We carried the seeds to The Leaf mashed in the tread of our hiking shoes, ferrying it back from the island where every spring we spend a happy weekend cheering on the warblers as they make their way north. We stalk the yellowthroats and black-billed cuckoos just as the garlic-mustard seeds are popping from their silique capsules, then we march into our own woods, pimping with every step.

First, our Woodland Garden was dotted with this rather pretty, airy weed; then a growth bloomed around the compost bins; now a vanguard creeps into the forest, stealing light and water and soil from the wild ginger, the bloodroot and hepatica, the trilliums and jack-in-the-pulpit. No more morels. No more Dutchman's-breeches. No more trout lily. The woodland poppy is endangered; the pale mauve wood aster threatened. The butterflies, which I go to such lengths to attract, dine on the deadly crosses of the stinky garlic-mustard blooms and fold up their wings to die. The deer are smarter: they refuse the free lunch. I am the plant's only predator.

So I awaken on this spring day thinking not of resurrection but of death. There's not a moment to waste. Already the garlic mustard is setting buds. Another week and I'll be at the mercy of the curse in that lethal garden rhyme: *One years' seeding, seven years' weeding.*

By 7 a.m. I am dressed and ready for battle. Atlas-grip gloves to give my fingers purchase on the slick, curvy necks of those devil mustards; freshly sharpened trowels; dark-green garbage bags to imprison the pulled plants until they rot; and vinegar. I've tried smothering the weedy growth under plastic mulch. I've tried burning it off with a propane torch. This year, I'll douse the soil with acetic acid to sear the seed and slow the growth of those I fail to remove.

The sun is a red ball rising above the gauntlet of cloud laid along the horizon. As I walk toward the trees, a ruddy light tips their still-bare branches. I advance up the path through the Woodland Garden, past the primrose and daffodils, past the pulmonaria with its tiny trumpets in two-toned baby pink and blue. Ahead of me, a low wall of green. I head toward it, into the woods, to the edge of the garlicky growth. This is the fire line I will hold, working my way back toward the house.

For a moment, I think of giving up. Surely this is a fool's errand. The ground between the trees is a blanket of fresh, bright green. How can I ever remove it all?

It's just a weed, I tell myself. A plant growing in the wrong place. Wrong, in my estimation. Exactly right, from the garlic mustard's perspective. See how it thrives. Who am I to chase it out of these woods?

I notice a wild honeysuckle blooming white against the maples. I pick a bud and suck it, the way I used to when I was a girl, walking to church. My hand strays to the garlic mustard. I nibble the edge of a leaf. It is tart, a tangy counterpoint to the honeysuckle, a sweet-and-sour morning salad. If I were Gargantua, I would down this bed of weeds for dinner.

> If I were Gargantua, I would down this bed of weeds for dinner.

Maybe the chickens would like it. Would it make their eggs taste of garlic?

If I left it alone, would it choke itself out? Given time, would another, more perfect weed, a white knight of a weed, come swooping in and return the woods to the ephemerals?

I set down my tin pail and pick up a trowel, plunge it into the soft earth to pry up a curvy root, which I then grasp and wrest from the ground.

Maybe, I think as I shove the plant into the garbage bag, I just want to be in charge.

THE SHIRLEY SHOW

I SHOULD HAVE PLANTED THE SHIRLEY POPPIES in the fall. "Just sprinkle the seed over the snow," the Garden Guru said. "November, March, April, it doesn't matter. They'll germinate when the light and warmth are right."

But who knows where I put the seed I collected last summer? I still can't find it by mid-April, after two warm spells have left the ground bare. The snow melted quickly this year—so fast a thick fog rose up from the ground. Vaporous snow. The Garden Guru told me to sprinkle the peppery seed over the last shreds of white stuff, but I've missed the moment.

"You can see how thickly you're planting," she said, "and the seeds love the snow juice."

Last year, I did manage to sow the poppy seed on snow, thanks to one of those freak storms that we think should stop in March but never do. Most Aprils bring at least one good blizzard. Not a dusting—we get those, too—but a real winter snowfall that lays down a mantle four, six, ten inches thick. It came while we were at the Hut, the log house in the woods where the Carpenter and the Garden Guru retreat from the world. We'd gone up to see the daffodils naturalized through the clearing around the cabin. They were glorious when we arrived; hours later, their trumpets were bowed under a heavy white shroud.

While I wait for a freak spring storm, I cruise the nurseries, but there is not a packet of Shirley poppy seed to be found. Why is that? Am I its only fan?

I like to think the plant is named for a Shirley: a dimpled Shirley like Shirley Temple, or a sweet Shirley like Shirley Jones of *The Partridge Family*, or a goofy, high-kicking Shirley like Shirley MacLaine, or the perky girly-girl Shirley who hung out with Laverne, or maybe that redheaded Anne Shirley, who lived in a gabled house.

But no, Shirley poppies are named for where they were born, in the parish of Shirley, England. They were coaxed into existence by the good and observant Vicar Wilks, who one day noticed a poppy in his garden, a volunteer that sprouted from seed no doubt blown in from the adjoining farm, where European wild field poppies, the so-called corn poppy, bloomed in profusion without any help from a gardener's hand.

Corn poppies grow everywhere in Europe. I've seen them waving in the gravel between train rails as lustily as between rows of corn. These bold little flowers were just beginning to bloom on that first Sunday in May 1915, when John McCrae wrote, "In Flanders fields, the poppies blow, between the crosses, row on row."

The corn poppy is small-faced and bright red, with a dark blotch at the base of the petals. The variant Vicar Wilks found in his garden in 1880 had petals rimmed in white. For years, he diligently collected seed and judiciously sowed, playing matchmaker to the poppies, until eventually he produced a distinct ornamental cultivar with petals in shades of white, lilac, pink, and red, their centres creamy and their edges blanched white, as if they've just pushed up through the snow.

Rummaging in my seed drawer for the flowers I usually plant around the time of the last frost—nasturtiums, bachelor buttons, zinnias—I finally come upon an envelope of pods. It's a haphazard affair, this collecting of seed. I keep coin packets handy on the cookbook shelf, and when I think of it, I stuff them with dried seed heads, casings, and siliques. I'm rarely diligent about labelling my stash. More often than not, I end up with unidentified heaps of granules and bits like fingernail parings that I toss into a mystery bed in early June. But the poppies are distinct. I crush the dull brown pods between my fingers: seed pours out.

Is it viable? A big question, one I ask myself often, not only in the garden.

Is it viable? A big question, one I ask myself often, not only in the garden.

I toss the poppy seed over the dark soil along one side of the Kitchen Garden, imagining a band of nodding red and pink and white guiding my eye up toward the woods. The seed disappears like pepper on oven-roasted beef. I toss on some more.

My moment of wild abandon over the bare spring soil exacts its price. I have forgotten, if I ever knew it, that poppy seed weighs in at two hundred thousand seeds per ounce. In my garden, they sprout thick as mesclun, the shoots growing plump despite the lack of snow juice. A bounty of Shirleys. My garden guides tell me I must thin the patch to at least three inches between plants. If I want them to rise above my knees, I'll have to give them a good hand-span of growing room. I yank out seedlings by the fistful, leaving polka-dot tufts that I winnow to a few spindly stems. In a week or two, I'll decide which of the trio will be allowed to live.

I don't like this job. I know what it is to be those crushed, discarded poppies, the ones that fail to make the cut. I, too, was never chosen for the team. *Only the strong survive. Sacrifices must be made.* I mutter the homilies under my breath. *It's a crappy job, but somebody has to do it.*

Or do they? Why exactly do I spend such long hours preparing this patch of soil? Sowing, thinning, then thinning again, feeding and weeding, pinching back to encourage bushy growth, waiting for that morning in early July when I walk up my Beloved's stone steps and there the poppies are, like a parade route lined with pink-and-red tissue-paper flags, glistening with dew, waving me toward the trees, where the leaves are sun-tipped a golden rose.

The day warms, drawing from the earth a breeze that sets the Shirleys to bobbing and twirling, filling the air with the lightest, sweetest scent, and the question dissipates, already answered.

I could cut them. Shirley poppies make good house-flowers if you pick them just before they open, while the green calyx is still attached. Sear the cut stems over a flame and they'll last in water for days. But I prefer to leave them where they grow, watching the nodding stems rise and stiffen to lift the bloom into the air, the petals spreading wide in the sun, folding like butterfly wings just as the finger-shadows of the pines reach across the lawn.

I have other poppies: the buttercup-yellow celandines (*Stylophorum diphyllum*) that have naturalized in the Woodland Garden; the self-sowing California poppies (*Eschscholzia californica*) that paint the colours of a Pacific sunset in my garden in late spring. Though I've tried, I've had no luck with the lovely blue Himalayan poppies (*Meconopsis betonicifolia*) that seem to grow only in gardening magazines. But it isn't a real poppy, anyway. None of them are.

The Shirleys are true poppies, members of the papaver family. (To be specific, *Papaver rhoeas*, which is the corn poppy, too.) When their stems are cut, they bleed a sticky white blood. The Orientals (*Papaver orientale*) are true poppies, too. I have a trio in the Forge Garden, that I grew from a root I dug up when I left my northern garden. They are planted in the lily patch, a bridge between the springtime 'Stella d'Oro' daylilies and the summer Asiatics. Behind them is a cascade of irises, though it's touch and go every year whether the blowsy China-red poppies will open while the Siberians are still screaming neon blue.

The opium poppies (*Papaver somniferum*) resist such orchestrations. They grow where they like, a perpetual surprise, their grey-blue stems lifting up flowers that might be thick and bold as pink carnations or a single, delicate mauve plate. I could be arrested for growing these in the United States. In Canada, they are simply registered as poisonous, a potential upset to the many stomachs of grazing cows. These are seeds I collect with diligence, throwing them into cakes and breads and tea. The heat supposedly destroys the active ingredient, but I wonder. Not long ago, on an episode of *MythBusters*, a person who ate four poppy-seed bagels was tested for narcotics. The results were positive.

I doubt I could drive the opium poppies out of my garden even if I wanted to, which I don't. None of the papavers is very demanding: they like lots of sun, not too much water, soil that's not too rich. A field, in other words. A railway bed. A burying ground. I appreciate their easygoing habit, even their copious germination, their rosettes of lobed and hairy leaves, their brilliant, throw-caution-to-the-wind bloom.

If I'm lucky, the Shirley show will last a week, maybe two. With a little dead-heading and just the right weather conditions, it might extend to three. Once, it was over in less than a day. Just as the blooms began to open, hail like an Old Testament malediction flattened the bed to a tangled heap. The devastation called to mind the young girl in the Sinclair Ross story "A Field of Wheat."

"Why did the beauty flash," she thinks to herself, "and the bony stalks remain?"

I push aside that vision as I pluck out the tiny seedlings and firm the soil over the roots of the Chosen Few. I see only the loveliness they will be, what I imagined as I first sprinkled the seed, those blushing pearl-rimmed poppies waving me into the woods.

ASKING THE EARTH TO SAY BEANS

I BOUGHT MY FIRST HOUSE FROM AN ELDERLY COUPLE who had raised their large family in that small shack in the woods of northern Ontario. My first husband and I arrived to view the house in early December, when snow covered the gardens where they grew all their food.

As we came through the door, the old lady was pulling delicious cinnamon-and-sugar pastries from the woodstove. My mother called them pinwheels, but this woman was from New Brunswick, of Acadian roots. "Nun's fart?" she said, passing the plate. On our second visit, the air in the kitchen was heavy with nutmeg and sweet grease from the doughnuts she was frying. The third time we met, the stove was cold. She was leaving the little house after thirty-five years, going back to the Maritimes. At the doorway, she leaned toward me with a parting word of advice. "Never plant potatoes before Rogation Day," she said.

Every year I think of this strange farewell. What, exactly, is a rogation? Not being a religious person, I didn't know enough to consult a priest. Instead, I headed for the dictionary.

The word comes from the Latin *rogare*, meaning to ask. Traditionally, these four days were devoted to asking for God's mercy, not for oneself but for all creation. The devout would walk in procession around the boundaries of their parish—"beating the bounds," they called it—as they blessed every tree, every plant, every seed.

There are two sets of Rogation Days on the Christian liturgical calendar. The first, the Major Rogation, falls on April 25, but there are three Minor Rogations, too, on the Monday, Tuesday, and Wednesday before Ascension Thursday, which comes forty days after Easter. This year, the Minor Rogations begin on the day we celebrate Queen Victoria's birthday, which is also the day that growers in my part of the country put in their gardens.

Just as the timing of the Rogation Days depends on Easter, the particular date of Victoria Day is determined by the need of Canadians for a long weekend, which has relocated the queen's birthday to the last Monday on or before May 24, the day she was actually born. This year, Victoria Day is May 18, which is as early as it gets.

Odd, then, that Victoria Day weekend, regardless of when it falls, always seems exactly right for planting. Maybe it's habit, the spring holiday provoking the horticultural urge just as Labour Day stimulates an impulse in me to buy notebooks and pens. Whatever it is, by mid-May the natural world has reached a tipping point. For weeks, the season has been swelling, and now conditions are poised for an explosion of growth. The woods are greening. The apple trees, like the dandelions, are in full, fragrant bloom. Every weed seed in the garden is sprouting up lush.

Planting, of course, is not seeding. I've been seeding for more than a month. Spinach and radish seed went into the ground as soon as the snow melted, and lettuce and beets not long after. In this gently unfolding spring, I've long since seeded all the herbs and greens and the hardier annual flowers like bachelor buttons, coreopsis, cosmos, sweet peas, and poppies.

A lot of the bulbs and tubers are in, too. The garlic was planted last October: it competes with the asparagus, putting on inches every day, like an athlete on steroids. The oxalises are poking up their first leaves, as are the dahlias. The gladiolas haven't pushed their noses out of the ground, but I know they're down there, getting ready.

The seedlings are about ready to burst out of their pots. On the first of May, the day the Farmer harrowed his fields, I moved them out from under lights into the sheltered brightness of the Garden Room. When the violets bloomed, I transplanted the tomatoes and zinnias and set them outside in the dappled light of the stone terrace, bringing them inside on cool nights. Now I tell them to brace themselves. I gather them in flats that I mass on the back stoop, together with my basket of tender seeds, pails of canna bulbs, and a sack of sprouting potatoes eager for the delicious dampness of the earth.

"Lots of work," my Beloved says, shaking his head.

But I love planting day. I want to get up at dawn and march around the periphery of The Leaf, blessing every brand-new leaf and seed, imploring the gods for equal parts sun and rain, mercy in the matter of earwigs, squash beetles, and slugs.

My garden could feed a family of ten. We're only two, but I plant it anyway, hauling the harvest to the Younger Son, the Elder Son, my Beloved's two Daughters. The Frisian takes some of the largesse, as does the man who comes on Saturdays for eggs. My Beloved wants to open a roadside vegetable stand. He has visions of long summer afternoons spent reading in the shade of a rough wooden cabana he's built himself.

"You'd have to sell things, " I remind him. "Like zucchini."

He winces at the memory of his mother's zucchini bread, zucchini soup, zucchini meat loaf. "Maybe we could take them to the city," he says, "check for unlocked cars, and stuff them in the back seat."

I drop bushel baskets of produce at the food bank instead.

I remember none of this excess on planting day. A kind of amnesia sets in as I build hills for the squash and cucumbers. The seeds are so small. Surely one zucchini plant is insufficient. Besides, they won't all sprout; the birds are voracious, and the bugs. I don't hold to folksy gospel, but still, as I plant, I sing the sower's rhyme:

"One for the rook, one for the crow, one to die, and one to grow."

The day passes swiftly. My Beloved opens bales of old hay and forks it over the garden until it rises well above my shins. To plant, I part it like the Red Sea, exposing a stream of damp, dark soil that my Beloved works with the Garden Claw. I press in the small red and yellow onion sets, the thin blades of leeks, heel in the red and green cabbages for winter storage and the Chinese cabbage for summer salads, the Brussels sprouts for late fall, the broccoli and cauliflower that take up too much space and never amount to much, though I always hope.

My Beloved cuts ten-foot sumac poles for the tomatoes, forty in all, divided roughly equally among the sauce tomatoes ('Sicilian Saucer' and 'Roma') and the eating tomatoes ('Brandywine' and 'Large Red'), with one red and one orange cherry-tomato plant nestled each inside its own wire cage.

He digs the trenches for the potatoes while I cut up last year's leftovers, an eye to each piece, which I lay out to "green" on newspapers spread in the Garden Room. I am not religious and I'm not superstitious or sentimental, either, but I'll wait, as the old lady directed, until after Ascension Thursday, the last of the Minor Rogations, to drop them in the ground.

My Beloved sets up the poles for the beans, a long teepee arrangement that by July will form a leafy wall across the back of the garden. On one side 'Cherokee Trail of Tears,' in remembrance of the long march of the Cherokee, and on the other French 'Emerite,' in remembrance of our dinners through the autumn we spent in Provence. I open the rows for the drying beans—red kidney, white cannellini, Mexican pinto, and navy—and for the green and yellow bush beans.

The seeds themselves will have to wait, for although I am not religious and I am not superstitious, I can still hear the old woman's voice in my ear:

"And don't plant your beans," she said with a wink, "until you can pull down your pants and feel the earth warm under your bum."

GARDEN SUMS

AT THE NURSERY, I SPOTTED A LONG SHELF of alchemillas, and thought, How gorgeous! Chartreuse leaves pleated like the ruffles under the chin of an Elizabethan queen. No wonder it's called lady's-mantle.

I considered the price and decided to buy one. No, three. Okay, I thought, I'll blow the budget and get five, congratulating myself as I made my way to the checkout that I had stuck to the planter's rule of shopping only in odd numbers.

At home, I set the young alchemillas in the Pine Garden between a handful of 'Plum Pudding' heuchera and a lacy green drape of *Lysimachia nummularia*, otherwise known as creeping Jenny.

Some might have mistaken this early garden at The Leaf for a species garden, the plants were set so far apart. In a true species garden, each plant is grown and enjoyed for its intrinsic value—a dying breed, a rare cultivar, a native that has long since lost its habitat. They're collectibles, distinct and unusual, not workhorses like coral bells and lady's-mantle and creeping Jenny.

"It's a bitsy garden," the Garden Guru declared when she stopped by for a visit.

I think of her as my mentor, my inexhaustible font of horti-cultural knowledge and lore. Usually, she is careful not to pro-nounce on my unschooled attempts.

"What's a bitsy garden?"

She laughed. "A bit of this, a bit of that."

"Just you wait," I said, a bit defensively.

The truth is, I don't like species gardens, or bitsy gardens, either. I gravitated toward *Alchemilla mollis* in the first place because there were dozens of plants on that nursery shelf, a long, crinkly swath of sturdy lemony green that drew my eye forward instead of locking it in place with a single bloom.

All my gardening life, I have wanted to grow in swaths. Is it any more unnatural than setting a plant from Mongolia, one from the Hebrides, and another from the steppes of Georgia to rub shoulders in a garden bed? Doesn't nature sow with a generous hand? Think of the drifts of Queen-Anne's-lace by the roadside, the carpet of Dutchman's-breeches in the woods.

But I have not always had the luxury of landscape. My tiny city garden backed onto an old quarry. The south and west boundaries were sixteen-foot limestone cliffs; the north, a high wooden fence. Manitoba maples overhung the property from above, so that the small enclosed yard was, in effect, a grotto—or would be, soon. I envisioned something like the famous fern grotto on the island of Kauai, where a waterfall trickles down the stone, bathing ferns in perpetual mist. A garage once occu-pied the very back of the yard: a mechanic's pit still gaped in the middle. I imagined a pond with koi swimming lazily in the filtered light.

I had big plans, but little cash. Then I discovered a farm close to the city that grew perennials: gardeners were encour-aged to bring buckets and a shovel to dig their own, five dollars apiece. I chose a large healthy hosta called 'Fire and Ice,' white leaves that would glow in my grotto, their edges deep green.

"Just one?" asked my Beloved.

"Remember the loaves and fishes?" I said.

I trained the hose on the hosta roots and gingerly wiggled apart the multitude of whorled noses that made up the plant. When I was finished, I had seventeen little bundles of 'Fire and Ice,' each with its own thin, bare root.

"Very small loaves," my Beloved noted.

"Just you wait," I said.

Those baby hostas looked lost that first year, pegs too small to hold up that vast sheet of stone. The second year, they filled out a little, and by the third summer, their foamy leaves flowed along the limestone to pool in the mechanic's pit.

At The Leaf, I have continued my frugal arithmetic, dividing wherever I can. My single white daylily, the spectacular *Hemerocallis japonica* that I purchased for more than I am willing to disclose, has reproduced so vigorously that it now flanks the curved stairway that leads down into the stone garden and rises in a bank behind the blood-red Asiatic lilies, too. The self-sowers are as welcome in my garden as the easily divided. The year we arrived at The Leaf, I bought a 'Johnson's Blue' geranium, unable to resist its striking mound of sapphire flowers. I collected the seed, thinking I'd start more, which I did, though I needn't have bothered. Every year, new seedlings cluster around the mother plants like chicks around a hen, until now swaths of blue geraniums spill into the meadow-field and throw themselves onto the grass.

By *swath* I do not mean row. I do not mean border. I do not mean a narrow rim of plants like a slash of coloured liner above the eye. I mean smudge. I mean sprawl.

Gertrude Jekyll did it best. The great garden designer started out as a painter, but her eyesight failed in mid-life and so she turned to plants. (She wrote all thirteen of her books after the age of fifty-six, an inspiration to late bloomers of every sort.) Jekyll is credited with bringing the roadside blooms she found in English cottage gardens—daisies, brown-eyed Susans, buttercups—into the formally designed beds of the great houses, reclaiming British horticultural heritage and planting it side by side with the ornamental exotics plant-hunters brought back from Asia, South America, and the Himalayas. She set out her plants in swaths, great bands of bloom that seemed laid on with a palette knife, bold strokes of colour that even she, in her near-blindness, could see.

I've never messed with paints, and my sight, for reasons I don't understand, improves with every year. Yet here I am, teasing the 'Plum Pudding' into dozens of small seedlings, arranging the offspring of my five lady's-mantle into a ten-foot drape, encouraging the creeping Jenny to roam with abandon past the hedge of 'Fire and Ice.' The wait has not been long. Already, goaded by the garden's mad mathematics, I am planting in broad strokes, with an English painter's eye.

FORGET-ME-NOT

FOR YEARS (AND PERHAPS STILL) the Alzheimer Society mailed out packets of forget-me-not seed as a heart-rending fund-raiser. My mother died of Alzheimer's, and so I scattered the seeds in the Hortus Familia, a bed of plants that reminds us of our families—a 'Blue Moon' balloon flower for my Beloved's father, who sang in a forties swing band under that same nickname; the golden spirea that we sisters gave our parents on their golden wedding anniversary, and that I dug out of their backyard after our father's funeral, hauling it home on a tarp spread over the back seat of our little car; the stones from the Newfoundland shore where my Beloved's mother was born, arranged near the red peony I released from under the eaves-trough drainpipe of her house when she died.

The forget-me-nots were for my mother: the blue of her eyes, the sweetness of her disposition, the way thoughts of her insinuate themselves, even now, into every hour of the day.

From that simple beginning, the forget-me-nots have cast themselves through every garden at The Leaf. A border by the cedars, a blue circlet around the apple tree, a soft haze out of which the lily-flowered tulips rise, bouquets scattered through the Forge Garden, a spray here and there in the Arbour Garden, and one solitary plant in the Shed Garden, where it has no right-ful place among the naturalized plantings of wild phlox and woodland poppy, yet there it is, as at home between the may-apple and the ferns as anywhere.

"What's that?" our neighbour the Fireman said one May afternoon, pointing to the hump of lawn over his septic tank, the grass thick with tiny blue flowers.

"Forget-me-nots," I said. "They run wild in my garden."

"No problem," he said. "I'll get rid of them. I'll spray."

I recoiled, as if he'd suggested weeding my brain-matter clear of my mother. Yanking out the memory of her making tiny doll sandwiches for the picnic-in-a-wagon that my Sister the Therapist and I entered in the Dominion Day parade. Withering the thought of her arranging the carnations my father bought for her every week. Crushing the low, curving bowl with the Japanese fisherman, whose line dangled among a pool of cacti. My mother wasn't aggressive like forget-me-nots, but you wanted her everywhere, always within sight, close enough to touch.

What grows in my yard is the biennial European wood-land forget-me-not, *Myosotis sylvatica*, so named because the leaf looked to Linnaeus like a mouse's ear. In many languages, though, its nickname has to do with remembering: in French *ne-m'oubliez-pas*, and in German *vergissmeinnicht*. My favourite story about forget-me-nots is a German folktale. A young man and his love are walking by the river Danube when he spies a swath of blue flowers. He scrambles down the bank to pick them, but the bank is steep and he loses his footing, falling into the deep waters. As he sinks from sight, he tosses the flowers onto the bank, calling out to his lady, "Forget me not!" This little blue, five-petalled flower with its golden centre evokes in me that same loving longing.

An almost identical flower (minus the yellow centre) rises high above the leaves of the false forget-me-not, *Brunnera macrophylla*. The brunnera is a much more civilized plant. Like the true forget-me-not, it flowers in spring, but the leaves are larger; they mound into a small shrub and stay fresh-looking all summer, whereas the other dies off by late June. Brunnera grows in shady spots, and comes in varieties distinguished by the markings on its heart-shaped leaves—dull green with a silvery pattern like crazed pottery ('Jack Frost'), green with a tracing of silver ('Langtrees'), or pale green with a rich creamy centre ('Hadspen Cream').

This spring, my 'Hadspen Cream' brunnera failed to appear. I have no idea why. The winter was not particularly harsh and the plant is supposed to be hardy to zone 4. Yet where I am pretty sure it was supposed sprout, there is nothing but a slimy black stem.

All around it, the true forget-me-nots bloom in profusion. I haven't seeded for years, and I pull up the plants religiously at the end of every May, yet here they are each spring in some new place, as unexpected, as persistent as memory.

HOME-GROUND INVASION

It was during my third spring at The Leaf that the Rosarian and the Humanist invited me to the annual village plant sale. "Bring a wagon," they said. "And boxes."

The village is a ten-minute drive down our old carrying road. We arrived shortly after seven-thirty on a fresh Saturday morning. This was only the second year of the sale, but word had spread. A knot of people was already gathered at the end of a long lane, sipping coffee and nibbling on muffins. More streamed down the main street, pulling wagons or pushing wheelbarrows laden with cardboard boxes. Ahead of us, a wide, generous lawn was parked with half a dozen hay wagons, each alive with hundreds of potted plants.

"Where do they come from?" I asked.

"All the best gardens," the Humanist answered, nodding sagely.

"Are some of them yours?"

The Rosarian started at the vision of others trampling among his darlings. "Heavens, no!"

By eight o'clock, more than a hundred of us were surging like cows at milking time against the sign that said PLANT SALE OPENS AT 8 SHARP. I wished I'd brought binoculars. Gardeners all around me were bending into a runner's stance, ready to make a dash for the prize they'd spied.

I brought home dozens of plants. These were early days at The Leaf. I'd spent weeks ripping off sod and digging out the roots of staghorn sumac that snaked through the soil, sending up frilly tufts that threatened to become trees. I pulled on their root-tails and followed them ten, twenty, thirty feet across the bed until I thought I must surely have got them all. More than anything that spring, I wanted beds full of flowers. Brown-eyed Susans, phlox of every colour, perennial geraniums, Japanese anemones, mother-of-thyme, various sedums, pulmonaria, blue flags, bleeding hearts, mountain cornflower, forget-me-nots.

In no time, they spread themselves out, casting their seed, extending their roots and rhizomes, making themselves so at home that now I'm afraid I'll never get rid of them.

I have no one to blame but myself. I invited these sprawling profligates in. What else, I think now, did I expect to find at a plant sale but the plants a gardener is most happy to be rid of?

When the tag on a nursery plant reads SELF-SEEDER or SPREADS EASILY, gardeners beware. It should say HOME-GROUND INVADER. IMPOSSIBLE TO ROOT OUT.

The beds that seem so sedate in April, and maybe even May, spiral out of control in June. The self-seeders are getting it on like teenagers home alone. Every bit of bare ground bristles with seedlings—baby hostas, sprouting borage, little rosettes of pink malva, brown-eyed Susans, forget-me-nots, and the magnificent, insidious *Impatiens grandiflora,* whose innocent-looking sprouts tower to a gorgeous eight-foot hedge flush with soft pink. If I get at the sprouts soon enough, I can slice them off at the roots with my trusty loop-hoe. More often than not, though, they are already too fleshy by the time I notice, and they force me to my knees, ripping them out of the earth like handfuls of hair.

Still, in the same way a person tolerates the bad habits of someone they like, I don't mind some of these messy, high-maintenance plants. The low-growing sedums, for instance, are aggressive spreaders, but they're easily kept in line. Every so often I reach under the edges and rip out some of the growth, reducing the clump to a more manageable shape, giving some breathing room to the plants on either side. The same walk-by pruning works for most of the creepers: the mauve-and-white creeping phlox, perennial baby's-breath, mother-of-thyme, even the pushy creeping Jenny.

I have to be more aggressive with the lily-of-the-valley, which for all the delicacy of its bloom is as tough as an old queen. When I planted it beside the driveway at the city house, it pushed up through the asphalt. My Beloved planted it here at The Leaf, in the Arbour Garden, which was once a pile of old lumber and rumpled eavestrough and is now a leafy band of shrubs, trees, and Virginia creeper. He planted it for me, knowing how much I love the scent.

"You take care of it, then," I grumbled, an ungracious lover. I knew what would happen. Before the summer was out, the lily-of-the-valley was thick as thieves, sending its pernicious roots into the arbour, the lawn, and through his neat wood-chip path. Now, every spring, I cut into it with my spade and push it back several feet, confining it to its bed.

I knew about lily-of-the-valley, but the Japanese anemone took me by surprise. I've always admired its thick, peony-like foliage, the pale mauve flowers that sway in the breeze on those tall, tensile stems. I bought three at the plant sale that first year, and set them in a bed between the 'Stella d'Oro' daylilies and the 'Purple Mound' chrysanthemums, part of my endless quest for continuous bloom from the last snow to the first. For two years, the anemones sat there looking glum. I bought more, despairing of ever seeing blooms. Then, suddenly, they were everywhere, insinuating themselves among the daylilies, choking out the chrysanthemums, threatening the Kentucky coffeetree I'd finally planted in their midst, having given up on the anemones altogether.

"I should pot some up for the plant sale," I say to my Beloved.

"Good idea!" he replies, not catching the sarcasm. It is he who rescues every tree seedling the squirrels plant, lining them up along the back of the Winter Garden: oaks and walnuts and red pines. This year, as I pulled fledgling ornamental cherry trees out of the Forge Garden by the handful, he was carefully digging up Norway maple seedlings from the path, setting them into pots.

"For the roadside stand," he says, only half joking.

"You'll have to put GROW AT YOUR OWN RISK on every one," I warn.

The self-seeders, I can manage. Even the spreading ground-covers are tolerable. But I am undone by the phlox, whose roots sink halfway to the molten core of the earth, it seems. I've tried digging them up and still, like guests who can't take a hint, there they are, back again next year. Yet I love those bright pink and mauve and white blooms, and what else so graciously survives the hot drought of midsummer?

I think I may have found a solution, a way to keep the phlox in their beds. Last fall, I asked my Beloved to cut the bottoms out of three of the big, black plastic tree planters that have been piling up behind the shed. Then he dug deep holes in the west perennial bed and sank the pots in place.

"I bet the phlox will find a way out," he says with a twinkle.

I get the feeling he's on their side.

"Maybe, but this will slow them down."

I still go the village plant sale every spring, and we still stop at roadside stands to look over the pots of phlox and forget-me-nots and Japanese anemones for sale for just a dollar. When a friend offers me a small-flowered white anemone, I decline, even though I can imagine the blooms twinkling like stars in the darkness of the Woodland Garden. At nurseries I've learned to be cautious. I read the fine print on the labels. I've become adept at the language of invaders.

"We're rich!" my Beloved exclaims one afternoon, calling to me from the back of a nursery, where trees and shrubs are lined up against a chain-link fence. He points to one corner.

There, in big black pots, are row upon row of staghorn sumac.

"Thirty-five dollars each," he says, holding up the label. "And we have thousands! Next time we come into town, we'll load up a flatbed and make a trade."

Or maybe I'll just take them to the plant sale.

WONDER

My earliest years were spent on the fringe of a small village in southwestern Ontario. At the back of our lot, a split-rail fence divided my father's neatly mowed grass from a small, damp wood that thrived in the floodlands of the River Nith. The swampy wood is mostly gone now, which hardly matters, since it always seemed like an imagined place to me, like something out of a book, though not just any book. It was the forest inhabited by the Girl of the Limberlost.

Not long ago, I was speaking to ten-year-olds at a writing conference. When I mentioned *A Girl of the Limberlost*, the front row of little girls let out a collective, exuberant "Ooh!" I wonder if Gene Stratton-Porter knew what she was starting when she published her book in 1909, if she ever imagined that a hundred years later, girls would still be entering that magical literary forest.

Grown women, too. There are no swampy patches on the sloping shelf of limestone that is The Leaf, but there is a slightly damp pocket at the edge of the woods between the composting yard and the shed. This is where, in our third year, my Beloved discovered a wild honeysuckle. He released the shrub from a bramble of wild grapes, bumbleberries, and Manitoba maple saplings, carved a snaking path and laid a half-moon mosaic of stone where he set a simple cedar bench so that every year, in early June, we can sit in the waft of the honeysuckle's sweet perfume.

A path demands a garden, and so under the oversheltering elms and basswoods, I planted clumps of wild columbine within a border of chartreuse-green sedge that I found in the woods. I brought ferns from a neighbour's forest and wild phlox from the wood of another. Every spring My Beloved and I travel to an island in Lake Erie to welcome the returning warblers, stopping on the return trip at a native-plant nursery where the knowledgeable and amenable owner loads up our car with woodland poppies, blazing star, uvularia, and other native species dug from his own backwoods. I plant them among four species of my own native violets, then move in trilliums, mayapple, and bloodroot, too. I encourage the grapevines to drape like llianas from the trees.

It's all fake, of course. Manufactured wildness. Only the honeysuckle rightfully belongs here. And the jack-in-the-pulpit.

The path, as my My Beloved envisioned it, started at the back corner of the shed and wound between two trees toward the honeysuckle. Into the middle of his proposed path jutted the thick chocolate-brown noses of a nest of newborn jack-in-the-pulpits.

"Wait!" I exclaimed, reaching out to stay his shovel. I pulled out my Japanese knife, quickly pried up the jacks, and set them to the side.

The next spring, three more noses thrust up through the wood chips and nylon webbing of the path. I moved them to the small, damp garden in the ell where the kitchen addition meets the stone house, planting them in a space bounded by my never-blooming rhododendron, a bank of hostas, and a lush Korean lilac.

The jack-in-the-pulpits did not mind the move. In fact, they loved it. The ones I remember from my childhood brushed my ankles. I would have to squat to lift the stripey "pulpit" to see the velvety brown "jack" inside. The ones in my garden rise to my chest, a good four feet at least. Jack and the Giant Pulpit. What seems to be the flower of *Arisaema triphyllum*—the pulpit—is actually a spathe that wraps around the jack and flaps over it like an awning. The jack itself is a spadix covered with tiny, invisible flowers, mostly male in young plants, increasingly female as the plants mature. As the summer progresses, the spathe will wither and the spadix metamorphose into a magnificent shaft beaded with bright red berries.

Spathe. Spadix. Metamorphosis. A vocabulary of wonder.

When I was a small girl, our everyday dishes were pale yellow with scalloped edges, dishes from the forties that I've since seen reproduced in trendy kitchenware shops. When company came for dinner, my mother would get out her good dishes—white china, with a band of chocolate and spring green at the edge, and in the centre a perfect, perpetual jack-in-the-pulpit.

My jacks are blooming now, rising up above the ferns and the phlox. More chocolate noses are thrusting up in the path. I move them deeper into the native garden, into my own little Limberlost, feeling like a kid again, full of wonder and expectation, as if someone special is coming to dinner, or I'm about to settle into the pages of a much-loved book.

LIVING COLOUR

I HAVE NEVER UNDERSTOOD THE POINT of a white garden. Vita Sackville-West planted one at Sissinghurst in 1950, "an experiment which I ardently hope may be successful, though I doubt it." She wrote about her monochromatic plan in a January gardening column in the *Observer*, that venerable British Sunday newspaper. So benign is the English climate that on that January day, she was just about to set out her first garden plants. In the column, she imagines walking up a grey flagstone path to sit on a bench with her back to a yew hedge, looking out over a sea of grey-green artemisia, cotton lavender, and Saviour's-flannel (a name I have since adopted, preferring it to "donkey's-ears"). She saw spires of white iris, white peony, white Regal lilies (*Lilium regale*) rising from the pale bed.

"It is amusing to make one-colour gardens," Sackville-West writes, rather weakly, I think. The tone sounds defensive to me.

Gertrude Jekyll made one-colour gardens, too. Blue gardens. Gold gardens. I don't think she ever made a white garden, though I doubt that was because she didn't think of it. In the 1950s of Sackville-West, white was virginal, the colour of innocence and bridal gowns. In Jekyll's Victorian England, white was the colour of death.

I can appreciate the challenge of an all-white garden, of finding a full range of albino-flowering plants. Still, the idea strikes me as odd, almost aberrant. To strip Nature of colour seems perverse. Our first winter at The Leaf, when I was dreaming of gardens, I bought a design book with a wheel that condensed colour to a dozen sharp wedges confined in a tight circle. It made me weep. In real life, colour is an endless spectrum that splits and splits again into a thousand million hues. Take the daylilies that spill like a waterfall from the Woodland Garden. They are just beginning to open, but when they are in full flood, the slope will be a shimmering yellow, dozens of shades of lemon, chrome, ochre, sulphur, and sun—such a multiplicity of tone that it seems an insult to call them merely "yellow."

Looking out over the pink, blue, mauve, and lavender of the June garden, I hardly see two tones alike. Such profligacy of colour! If we were talking about a person's wardrobe, I would put it down to vanity, or maybe aesthetics. But this riot of polychromicity has nothing to do with human pleasure. It is purely practical, a plant's way of advertising for pollinating insects and birds, like the neon sign outside a strip joint or the hot-pink tube top a young woman wears.

I have a Chinese-red dressing-gown I sometimes wear in the mornings when I pick the day's lettuce or a handful of chamomile for tea. The hummingbirds make a beeline for the enormous red flower that is me, hovering within inches of my shoulder, my elbow, my chest, flapping frantically. It's no accident that I see them most often feeding on the hot-pink coral bells, the lipstick-red mandevilla vine, the scarlet geraniums.

Birds in general are attracted to red or yellow flowers. Bees, on the other hand, can't see red. My Sister the Therapist, performed a study that determined even newborn bees show a clear preference for yellow and blue. Moths come out mostly at night, so the flowers they pollinate are pale or white, blanched hues that show in the dark. Beetles and bats go for white flowers, too. On a moonless night, though, even white isn't much of a beacon, which explains why night-bloomers tend to give off a strong scent, too.

What does it mean, then, when a flower changes colour? The Rosarian showed me a rose this week that blooms almost red, then shifts to sunset hues, before ending up a pallid pink. From a distance, it looks as if four species are grafted onto a single shrub. What sort of fickle pollinator has this evolved to please?

In my own garden, I grow pulmonaria, even though it is an awful self-seeder, because I am intrigued by the pink and blue flowers that bloom, both at once. I assumed it was simply a two-toned plant, indecisive, like pregnant parents who stock up with baby clothes, half in blue, half in pink. But when I came back from the Rosarian's rosebed, I peered at the last of the pulmonaria blooms and discovered that the freshest flowers were pink, while the ones on their last legs were blue. It's not a pink-and-blue flowered plant: every flower turns from pink to blue as it ages.

I've read about this colour shift in other species. There's a forget-me-not, *Myosotis versicolor*, whose flowers change from pale yellow to dark blue. The flower of the Siberian pea, *Lathyrus vernus*, opens red and then turns bluish green. The change of colour is an early-warning sign of aging, visible long before a petal wilts, an indication that a plant is about to release its pollen. It's like my hair, which turned white while I was still in my thirties, long before my skin began to hold its creases.

To change colour in the course of a lifetime is one thing. To do it in the course of a day is nothing short of amazing.

Even more remarkable are the flowers that change and then change back again. *Desmodium setigerum* is a climbing perennial with tiny lilac flowers that grows in Zimbabwe. When a bee lands on the flower, its weight "trips" one of the petals and the flower's reproductive parts burst into view. After the bee leaves, the flower's top petal falls down like a CLOSED sign, covering the anthers and stamen. Over the next hour or so, the petals turn from lilac to white then turquoise, a signal to other bees that someone else got there first. But that's not all. If the plant hasn't received enough pollen in that first visit, the colour reverses from turquoise back to lilac, the equivalent of flipping the sign on the door from CLOSED back to OPEN. In all of nature, only flowers have this ability to change colour and change back again. It reminds me of the genetically altered males in Margaret Atwood's novel *Oryx and Crake,* whose penises turned bright blue when aroused. When the message is in living colour, there can be no mistake.

> My garden is not merely colourful: it is scrawled with rainbow love-notes.

This language of colour in flowers is exquisitely refined. Take the *Aster vimineus*. Its flowers have a central disk of florets that are either red or yellow. The yellow florets contain more pollen than the red ones. So why not just produce all-yellow florets? Because the red petals are more attractive to birds. The red-centred flowers act as decoys to draw in the pollinators, which then pretty much ignore the red blooms and head for the pollen-rich yellows.

My garden is not merely colourful: it is scrawled with rainbow love-notes, each in its own distinctive hue.

I like to think of someone dropping a scarlet tulip bulb or a purple Johnny-jump-up seed in Vita's all-white garden. What subversion! This is exactly what Christopher Lloyd, that eminent and irreverent British horticulturist, recommends. "Two colours may be shouting at each other, but they are shouting for joy," he says. Experiment! he insists. Search out violent contrasts!

He is talking aesthetics, not practicalities. He isn't thinking of the birds and the bees, though they'd be happier with the shouting-match, too, than with a monotonous monochrome conversation.

A few days ago I visited the Rosarian and the Humanist. "The roses and peonies are in full flood. You must come before the week is done," insisted the Rosarian when we met on the road, him with his Scottie and me in my garden togs. I arrived at their stone cottage to the sound of the Humanist raking the paths in softly rattling stony sweeps.

The rose garden was indeed at its peak, the perennial border, too, with mounds of pink, white, mauve, and fuchsia peonies, a spring colour symphony. In the middle, a low orange lily blared its trumpet.

"We had visitors on the weekend and a woman said, 'Oh, you must rip that out!'" the Rosarian laughs. "I didn't tell her, but I like that flash of orange." He lifts his arm in the direction of the rose garden, the peonies. "There are colours that are soothing, but this," he says, coming to rest on the brilliant splash of orange, "this does something for you. Shakes you up. Gives you a jolt. I love it."

"I wonder how much Vita Sackville-West loved her all-white garden," I say.

"She didn't plant it because she liked white," the Rosarian declares, setting me straight. "She made a white garden so that her sons could find their way to their quarters at night across the courtyard of that big old castle at Sissinghurst."

Oh, I think, she had a *purpose*. Somehow that changes everything.

I have started a white garden of my own, just a small one, where the ell of the west perennial bed turns to stretch alongside the Garden Room. I'm planting it with Regal lilies, nicotiana, August lilies, and *Datura* 'Belle Blanche.' It's the fragrance of these night-bloomers I'm after, perfume to waft over us in the evenings when we sit out to watch the fireflies. We'll wait for the night-flying moths, too, and insects drawn not to colour but to its absence, and I'll think of Vita, making do with monochromes, and of her little boys, finding their way in the dark.

FULL BLOOM

A MEANS OF SUPPORT

IT HAS BEEN RAINING FOR DAYS. The peonies lean their heads on the grass like drenched terriers. The delphiniums flail, then snap with the strain. The tradescantias lie flattened in defeat.

It happens every summer. Flowers cradle rain like nectar; stems bend under the burden of water, droop and break. After a downpour, the garden lies in ruins.

"Can't you prop them up somehow?" my Beloved says, surveying the tangled mess.

I have tried. Every spring, I eye the commercial plant-props with the determination of a military mother. I must support my plants! I'm obviously not alone in my resolution. The walls of nurseries hang with dozens of variations on the theme: wavy wands in brass, aluminum, and plastic-coated steel; the Peony Pal, which looks like one of those mosquito coils we used to burn to keep bugs out of the old Westfalia; a plant support that could be a recycled barbecue grill on legs.

I am a Libra. I don't do well with too much choice. I ponder the options, make lists of pros and cons. That there are so many variations makes me think none of them will work very well. And they are expensive. Even the cheapest costs close to ten dollars, and I can't begin to count how many I'll need.

Mostly, I want support for the peonies. When we arrived at The Leaf, there were three ancient nests set in a row perpendicular to the house, bisecting the West Yard for no good reason that I could see.

"Why would anybody do that?" I wondered aloud.

"Why does anybody do anything?" my Beloved said.

Our first act of gardening was the transplanting of these peonies. We repositioned them to form a hedge parallel to the house, creating one wall of the "room" I envisioned, a wall that is a dark, waxy green from May to October, but flounders under the weight of flowers in June.

Every spring I leave the stores empty-handed and cobble together my own peony supports. I poke my mother-in-law's rusted tomato cages into the soil over the rising peonies, then stretch chicken wire (too narrow, really) or Paige fencing wire (too wide, really) across the top for the stems to grow up through. It works, after a fashion, though it is an eyesore during the weeks it takes for the peonies to bush out, and a booby trap to dismantle in the fall.

"Something more architectural, perhaps?" my Beloved suggests.

He leads me to the rusted iron bedstead we found among the sumacs when we cleared the stone terrace, then to the harrow he hauled out of the woods a couple of springs ago. A gardener we know sets bits of metal among his plants, creating firm ledges for them to lean against. The effect is surprising and delightful, something to do with the contrast between heavy metal and the delicate green of the leaves. But when I prop the bedstead among the anchusas, it looks like what it is—a piece of junk.

"Be whimsical, creative, even elegant," the gardening magazines exhort. They run photographs showing roses climbing over an iron bedstead(!), clematis making its way up a wooden ladder.

I'm a hoarder—I have plenty of old ladders and bedsteads and washtubs, but they look silly in my garden. The truth is, I prefer a more natural look. Once, digging the Woodland Garden, I found the jawbone of a cow, which I propped near the path, where a yellow groundcover overtook the prodigious molars. Every time I passed I thought of Jason and the dragon's teeth he planted and the warriors that sprang from them. That's the kind of support I like: one with a story.

In the Winter Garden, my Beloved erects eight-foot sumac poles that he cuts each year in the woods. I tie the tomato plants to them with raffia that I buy in a shank and fix to the top of one of the poles, my Rapunzel of the Brandywines. At the back of the garden, he builds a long, extended lean-to of poles for the 'Cherokee Trail of Tears' black beans, distant offspring of the beans the Cherokee carried with them when they were removed by long march from their ancient homelands in the American South to Oklahoma. In July, the Grand Girls play house inside the leafy bower like a couple of Gretels, and in January, from the kitchen, I watch as the shadows of the vine-twisted stakes slither across the snow, like snakes escaping from the garden.

Every three years or so, we make a tall, slender obelisk near the Egyptian onions at the entrance to the Kitchen Garden, leaning four sturdy hardwood poles together and wrapping them with wild grapevines that I tie in a vaguely heart-shaped knot at the top. Clematis grow up the vines, a different colour set at each side of the base, so they twine together as they grow, a wondrous, impossible sun pillar in white, pink, and mauve.

"The Victorians supported all their shrubs with willow wound around hardwood stakes," the Rosarian tells me. He has a willow grove, but we don't, and so last year, when we cut back the grapevines in the spring, I set some aside. I stuck pointed sticks into the ground around the delphiniums, the roses, and the peonies, then twined grapevines around the stakes, criss-crossing through the centre to create spaces for the rising stems.

It worked, it was beautiful, and it cost nothing. But I didn't get around to making my grapevine props this year, and so my peonies languish, leaning their heads on the grass.

"This is the way peonies are supposed to be," the Rosarian says, gently lifting one of my draping flower heads, then settling it back on the grass. He does not believe in staking peonies. He tells me about the peonies at the botanical garden near the university where he used to teach. The flowers fall like a fountain from the centre, he says. "Anything else just looks fake."

I look again at the peonies and see not a wet terrier but a graceful antebellum drape of pink pompoms, like the florets on a debutante's dress. I could, I think, become a convert. I warm to the idea. No more stakes, no more collars. Let the peonies go it alone! Let the delphiniums wave in the rain, throwing their water-weighted stalks over the leopard's bane. Why force them to be anything but what they are?

I wonder, Could I do this? Throw away my props? Learn to love disarray?

VEGETABLE LOVE

I HAVE ALWAYS THOUGHT OF PEAS AS TOO MUCH WORK: all that popping and thumbing of pods, and for what? A tiny pile of bright green pellets that are marvellous, yes, but are they worth the space the plants require, the time it takes to release them from their shells?

The answer for me was no, and then I discovered the 'Dwarf Grey Sugar.'

There are, of course, three kinds of peas: shelling peas, or English peas, where you throw away the pod and eat the little seeds inside; sugar snap peas, where you eat the pod and the developed peas both; and snow peas—also called sugar peas— where you eat the pod before the peas develop, and the leaves and the tips of the vine, too.

The 'Dwarf Grey Sugar' pea is one of the latter. The entire plant is edible, which immediately appeals to me. This particular sugar pea is an old variety, introduced to Europe in 1892, about the time the grandchildren were taking over this house that Squire Molton built. By modern standards, these plants are quite stunted, only two to three feet tall, but the leaves are a lovely purplish grey-green and the shrubs bloom in two-toned pinkish purple flowers that look for all the world like sweet peas.

The truth is, I grow Dwarf Greys as much for the flowers as for the pods.

I plant the peas as a flowering hedge down one side of the Kitchen Garden. For support, I build what I call a Newfoundland fence because that's where I first saw it, in my Beloved's mother's home village by the sea. It's a wattle fence, really—a series of straight twigs (I use sumac from our annual cut-back of this insistent invader) set at an angle in the ground like a row of backslashes overlapped with another set of backslashes going the opposite way. I make two fences about two feet apart, then broadcast the Dwarf Grey seeds in between, poking them into the soil at roughly two-inch intervals.

The irony of peas is that they grow best in cool weather but germinate best when the soil is warm. I try to take advantage of the inevitable warm spell in early April to get them into the ground, then hope for a cool spell to leaf them out. They aren't picky plants to grow. All through May and June, the Dwarf Grey hedge gives me pleasure, and then the peas come on in profusion. One cool, wet spring, we had peas on the table for ten weeks.

I learned how to cook peas from M.F.K. Fisher, the American writer who took food as her subject in the early years of the Second World War. What to do when the wolf is at the door? Cook it, she exclaimed, and wrote *How to Cook a Wolf* about living well on wartime rations. The chapter titles alone are worth the price of the book: "How to Distribute Your Virtue." "How to Be Cheerful Though Starving." "How to Be Content with a Vegetable Love."

It is in *The Measure of Her Powers* that Fisher describes the proper way to cook peas. First, the peas must be very green and gathered just before eating. Fresh-killed, I like to say. For Dwarf

Greys, that means pods two to three inches long, with a faint, just-pregnant bulge of peas barely visible inside. Fisher podded her peas; I slice mine at an angle or leave them whole. In a large, flat skillet, I bring a scant half inch of water to a boil, slide in the peas, and slap on a lid, then shake the pot like mad over a high flame. When the steaming stops, I throw in a big pat of butter, turn down the heat, and swirl the skillet for a minute, then ready my fork for a feast.

The pods stand out nicely from the Dwarf Grey plant, which makes picking easy. Even so, I always miss a few, but that is a good thing. Hidden among the foliage, the pods plump up nicely and dry on the vine, where I find them when I pull up the plants, the seeds dry and ready for next year's crop.

I've been growing Dwarf Greys now for twenty years from a pack I bought for a dollar in 1989. With luck, I'll never have to buy another.

SUMMER SOLSTICE

FOR YEARS, I REFUSED TO WEAR A WATCH.

"Time is a human construct," I'd tell anyone who asked. "Do the trees care if it's ten o'clock? Will the flowers wilt if they know it's five past six?"

Nature, I would say, has an inbred clock of its own. Flowers are its timepiece.

Every gardener knows this. We rise to the open trumpets of *Datura* 'Belle Blanche' and watch with sadness as they fade through the day, waiting for the evening, when the nicotiana lets loose its scent. We contrive to stroll past the oxalises as the first rays open its leaves and linger by the primrose at dusk, waiting for the bright yellow blooms to unfurl to the night moths.

Linnaeus, that great name-obsessed botanist, took things a step further. In his *Philosophia Botanica,* published in 1751, he describes a flower clock made by arranging plants in a circular bed according to when their blooms open and close. No one knows for sure if he ever actually planted an *horologium florae*— a floral clock—though he lists the plants needed to mark a full day's sweep of time.

It makes me wonder: what kind of a flower clock could I make? I'd start with goatsbeard, which opens at three in the morning, an hour that is neither morning nor night. At four, opening to first light is the rough hawkbit, a hairy native that I see along the roadsides when I take my morning walk, its orange flowers bobbing on tall, spindly stems. At five, the dandelion spreads its petals and so does the blue chicory. At six, pulmonaria. At seven, red hawksbeard, which looks an awful lot like hawkbit. Eight o'clock is a problem, but at nine maybe garden pinks or mallow—I'd have to check. At ten, according to Linnaeus's list, garden lettuce opens. That seems odd. A clock with lettuce flowers as a marker would be good only for a week or two in August. The middle of the day is a puzzle, too, one that resolves itself as the light wanes. Calendula closes at three, sweet alyssum at four, Icelandic poppy at seven, daylilies at eight, and at nine, if it's getting dark, the petals of the moonflower start to unfurl.

But what if it's raining? What if the clouds are hanging low? What if it is September instead of June? Nature may be precise, but it is contingent.

In *Philosophia Botanica*, Linnaeus defines three kinds of flowers: the Meteorici, such as oxalis and Japanese peonies, which vary their opening and closing times with the weather; the Tropici, like the misnamed four o'clock, which change their habits according to the length of the day; and the Aequinoctales, the true-blue timekeepers like the hawkbits and dandelions, which are steadfast in their habits, never varying the moment when their petals unfurl, no matter what.

Linnaeus allowed only Aequinoctales in his clock. At The Leaf, we're more accommodating.

"The oxalises are folding up their leaves," I say to my Beloved. "I'd better start supper."

"Not yet," he replies. "The four o'clocks are opening. It's happy hour."

The garden is not only my clock but my calendar, too. I don't have to open iCal to know that spring is over, summer has begun. The columbines are fading. The tulip leaves have wilted; the bulbs willingly let go their stalks. The ground under the late Preston lilac is pale with mauve confetti. The forget-me-nots are a haze of propagating grey. My early planting of lettuce has grown tough and bitter. The blue fescue has gone to seed. The Farmer is taking off his first cut of hay. The strawberries are ripening, the peas are flowering, the chamomile is ready to pick. The peonies are heavy on their bushes, and this morning, the first bloom opened on the 'Harison's Yellow,' a rose the Rosarian gave to me because it is almost as old as our house. It was discovered, he told me, on a farm in Manhattan before skyscrapers spread over that island like tall, bristling weeds. 'Harison's Yellow' is the sunny little rose that pioneer women dug up and took west, the history of American settlement told in a trail of roses, just as the story of early Canada can be read in the lilacs clustered where log cabins once stood, at the edge of abandoned fields— flowers marking time in centuries.

Digging in the dirt is not a young person's sport.

Lately, a subtle colour shift has been under way in the garden, from the soft pinks and mauves and shades of white of early spring to the hot poppies, neon-blue Siberian irises, and DayGlo lilies of summer. Delicate chartreuse leaves, translucent as the lime Life Savers I used to suck to a sliver, are giving way to tougher, waxier greens and the silver-grey leaves that reflect light and hold moisture through the parched dog days of July.

In a garden, even in June, you can see the seeds of its end.

Maybe that's why digging in the dirt is not a young person's sport. People come to the garden, my Sister the Therapist and

I agree, at the same time they come to the psychotherapist's chair: when they reach the halfway point, when the number of years that stretch ahead are no more than what's behind. The summer solstice of a life.

I have started to wear a watch. It's not that I've become a slave to time, but I've grown to like the noting of the hours as they pass. It's something I learned in the garden: a glimpse of the end turns the mind to beginnings. Seed to flower to seed, this year, next year, last year, the year before. The great life-measuring clock of the garden on full and constant view.

"What would you say to just one more garden?" I ask my Beloved on the first day of summer. "Perhaps a flower clock . . ."

ORDER OF THE DAY

ONLY THE ROSARIAN IS UP THIS EARLY, before the sun has fully cleared the horizon. He walks the road in the pale light, led by his Scottie, Clio, snorting and snuffling through the grasses. The dog is slung low to the ground, black as a 'Queen-of-the-Night' tulip.

This morning I wave from the Woodland Garden as they pass in front of the driveway. We keep an old, white-painted Muskoka chair on a limestone ledge there, in case my Beloved or I ever feels the urge to sit and admire the view, which is lovely—a long slope down to the creek where the bittern gulps through the dawn, then up the Farmer's wheat- and hay-fields to a fringe of bush against the southern sky. I've never sat in the chair, though I often gaze at it from wherever I'm working and smile, thinking of the possibilities.

The Rosarian chances to look up in that split-second when the view from the road is clear to the woods. He waves and turns toward the driveway, tugging at Clio's leash.

"I can't see your gardens from the road anymore," he grumbles as he walks up the lawn, past the cedars, the snowball spirea, and the *Rosa glauca* (a gift from him) that screen our gravel parking nook.

That was the point, I want to reply. When we arrived at The Leaf, the lawns gaped onto the asphalt.

"People can see us!" my Beloved exclaimed. He was used to the cabin we'd been renting, a log structure tucked deep in the woods, no one for miles.

"We'll plant hedges—and shrubs. Our house will be like Snow White's castle. No one will know it's there, even though it's just ten feet from the road."

For a decade we've been madly setting out cedar hedges, lilac hedges, and Arctic willow hedges; mixed hedges of blue spruce, Mountbatten juniper, and red and white pine; shrubberies of flouncy dogwoods, mock orange, a massive ninebark, and viburnums of various sorts; *allées* of oak, maple, and linden; and a linebacker of a weeping willow that stands guard on the eastern flank of the property, daring the neighbours to look.

At least my Beloved is happy.

"Gorgeous morning, isn't it?" I say to the Rosarian, ignoring his complaint.

He pulls out a blue-checkered handkerchief and wipes his brow. "It's going to be a hot one," he says. "We're wilting already."

"It's summer." I shrug, stating the obvious.

"Too hot to work outside."

If I had my way, every day would be exactly the same. I'd get up with the sun and go to my desk, where I would write until lunch, after which I'd go out to the garden, balancing mind-work with body-work until the evening, when I'd retire to my easy chair to read and make notes. In this imaginary life, I move in an ordered way through my days, going page to page, garden to garden just as my mother advanced through her week: washing on Monday, ironing on Tuesday, vacuuming and dusting on Wednesday, floors on Thursday, shopping on Friday, the weekend to relax.

This works, to a point, in spring, when the mornings are too cool, the plants too dew-laden for me to work outside comfortably. By the time I'm ready to garden, the air is warm and the soil dry. And by the time I'm ready to quit, the late-afternoon chill is driving me willingly indoors.

I keep an agenda, a red-leather book that opens flat with a red silk ribbon to mark the day. Everything that needs doing in the house, at my desk, in the garden, with our friends and family is set down in this book. Once a month, we see the Grand Girls. On Tuesdays my Beloved plays baseball (in summer) or hockey (in winter). On our anniversary we spend the day at the restaurant where we were married. On Easter and Thanksgiving, all four children come home with their families, if they can. In November, I order a new agenda, and in the week between Christmas and New Year's, I happily fill in its spaces with my favourite pen, setting down the birthdays, anniversaries, and events that will mark the coming year.

I like looking ahead, knowing where the handholds are. If I had my way, I'd walk the same paths, dawn to dusk, bed to desk to garden bed, a supplicant moving through her daily stations, no matter the time of year.

But summer is here. Everything has changed. By eleven in the morning, the sun is a weight on my back. By two in the afternoon, all the oxygen has been sucked from the air. I can't breathe, let alone work. Mad dogs, Englishmen, and my Beloved are still out in the midday sun. He's cutting the grass, his face glistening, his shirt soaked. I stay indoors, my eye on the bleached sheet of the sky.

For years, I resisted. I stuck to my routine. You could find me in the garden from two to five every day, regardless of the season. A hat was my only concession to the slaying sun, the flailing wind.

"You work too hard," my Beloved said one day.

"As hard as you," I replied. We were both a mess—mud-streaked, sweat-stained, slumped with exhaustion. "But it's not work. I like it. And it has to get done."

"If you're doing it because you have to, then it's work," he argued.

He likes to think it was that conversation that made me change my ways, but really, it was the heat—those summers when the mercury rose into the thirties and stayed there for six weeks. I bowed to the season, and left the house at dawn. Now, through July and early August, I work outside from five to eight in the morning, and again from five to eight at night, in that sliver of time between the cooling of the day and the stirring of the mosquitoes. I focus on the symmetry of it, my day in a new kind of balance.

"But it's five in the morning!" my Beloved moans as I roll out of bed and pull on my gardening pants.

"I have to get out there now," I whisper, kissing him back to sleep before I tiptoe away.

I spend the first days of summer tearing out the last of the forget-me-nots, pulling up the dried tulip stems, concealing the yellowed daffodil leaves or tying them into knots. I chop off the iris leaves at an angle, little grey-green arrows that will soon disappear under a drape of mother-of-thyme. The lupins have gone to seed: I have to get them out of the gardens. The alliums, too. The pulmonaria, columbine, and bleeding heart. Chop, chop, chop. I cut down the mountain bluet and the meadow rue so they'll bloom again in the fall, sweep aside the decaying leaves of the autumn crocus.

I rip out the spring plants with a gusto I haven't felt for weeks. I love this moment of transformation as the garden shifts from its soft spring face to summer, when the beds will be alive with searing yellows and reds—lilies, cleomes, late roses, gladiolas.

I like change in others. I avoid it in myself.

"Will you be working in the garden today?" I say to the Rosarian, by way of conversation. Though we can't see each other's gardens, on a still summer day I can often hear the Rosarian and the Humanist calling to each other across their yard. Not words, just the cadence of their sentences, a kind of music like birdsong, which I try to parse for meaning as an ornithologist might. I think I hear the Rosarian say, "Come in now. It's too hot," to which the Humanist replies, "Soon. In a minute. I'll just finish this." I wonder if they can hear the same chirping words exchanged across the beds between my Beloved and me. "Stop now." "Soon." "Come in." "When I finish."

The Rosarian and I stand on the open grass of the Croquet Lawn, the morning sun already blistering our skin. In Brazil, a woman said to me, "I can't step out the door. I turn to ashes." This day, even now at scarcely six o'clock in the morning, has that same feel to it. The dog is panting, tongue lolling, skeins of saliva silvering to the grass. I pat my secateurs and Japanese knife, secure in my pockets. I'm on my way to pick up the rest of my arsenal—wheelbarrow, spade, fork, and loop-hoe.

"No work for me today," the Rosarian says. "But you go ahead."

As we talk, the dog gets up, wanders over to the shade of the big white pine that towers over the terrace by the house, and plops itself down, leaving us to our conversation in the sharpening sun.

"Look at her," he says, with an admiration I crave. "She doesn't think about it; she just does what she has to do." He glances up at the white Muskoka chair. "I think I will, too."

LOVE PLANT

THE GRAND GIRLS CALL THEM BUTTERFLIES. First thing in the morning, when the leaves are folded upright, that's what these plants look like, a flock of big purple *Lepidoptera*. Then the light hits them and the leaves fall open, three magenta triangles with a wine-red stain where they join.

They look like shamrocks, but true shamrocks are trefoils, or clovers. St. Patrick, so the story goes, used the shamrock's three-leaved configuration to explain the Trinity to the Druids. Now it is sold as a symbol of luck around St. Patrick's Day, but because clover doesn't grow happily indoors, florists often substitute the three-leaved *Oxalis triangularis* as a remembrance of the Irish patron saint.

Oxalises aren't shamrocks; they are a member of the wood-sorrel family. There are over eight hundred varieties, most of them native to South America and South Africa. The one I grow is *Oxalis regnellii* ssp. *papilionaceae* (butterfly-like), a particular cultivar called 'Triangularis.' This large-leafed variety is edible—its leaves taste slightly sour, like sorrel—and it comes from Brazil. Oxalis, poinsettia, mandevilla vine—my few rooted remembrances of another life.

I've seen oxalises growing wild in the *bosque,* but never in any garden but mine. This is not a boast; I planted it by mistake. The packet of six corms was a gift from a friend who came to visit, and when she left, I stuck what looked like little pink pine cones into a bare spot of soil under the viburnum at the edge of the terrace. I didn't think to read the directions until later. INDOOR PLANT, APPROPRIATE FOR CONTAINERS.

"Well, let's see how you do out here in the wilds," I said. It was spring, and I had no time for redigging and repotting. Besides, if you love something, let it go free. Isn't that what the greeting cards say? I didn't know this plant at all, but that seemed a poor reason to deny it liberty.

The plants have thrived in their freedom. All summer they bloom, lifting small pinkish white trumpets high above their deep purple leaves. I have grown to love them very much, not least because their leaves are nyctinastic, one of those tongue-pleasing scientific words which in this case means they are sensitive to light. Cells in the joints detect blue wavelengths, causing the leaves to rise to a vertical "sleep" position when night approaches or clouds gather. The instant sunlight strikes, the leaves begin their descent, opening fully within minutes.

Red oxalis is supposed to be grown indoors in a bright window, but outdoors, in my garden, the colour deepens deliciously in shadowed light. I grow it as a border to a path through the lilacs, as a groundcover near the yellow roses, and spotted here and there amongst the hostas, where they raise their purple leaves in high clusters that look as much like flowers as the actual blooms.

I don't grow many plants that are so tender they need to be lifted and brought indoors for the winter, but I happily pull up the oxalises in late fall, along with the red-rose dahlia, the canna lily, and the gladiola bulbs. I dump the shallow-rooted clumps with their clods of soil into a big plastic planting pot, then cover the top with a pad of autumn leaves. The pot goes into the unheated crawlspace under the kitchen, where it passes the winter with the other pulled tubers.

One pot. That's all I'll need for next year. The rest of the oxalises get bagged and given to friends. In May, I loosen the soil a bit with my Garden Claw, tear the clods of oxalis apart, and shove the corms into the ground, finger deep. They like being crowded. They seem to figure out for themselves which way is up. Within a month, they are clumping. By now, in early July, they are in full bloom. Over the summer they'll spread into a deep purple haze studded with pale flowers that shine like glow-in-the-dark stars on a child's ceiling.

The first year I grew the Love Plant, as it is sometimes called, I became so enamoured with it that I potted several to keep in the house over the winter. But I couldn't stop thinking of them as hostages from the garden, prisoners locked behind the window glass, staring out at the sun and wind as they struggled to carry on, sending up straggly blooms above leaves that never folded up tight as butterflies.

After a few weeks, I emptied them into a plastic pot, covered them with damp newspapers, and shoved them under the kitchen with the others. Even plants, I decided, deserve a rest.

SEVEN YEARS' WEEDING

NATURE ABHORS NAKEDNESS. Scrape the soil bare and within a week it will be greening with weeds. Turn that soil over and you invite a plague of propagation that has been lying in wait for years.

The earth is full of seeds. The breeze, bees, everything that creeps and crawls, even simple gravity conspire to keep the planet green. Darwin knew this, but he was curious: how many weed seeds lay waiting in the soil, exactly? In the spring of 1857, he decided to find out. He opened a patch of earth thirty-two feet square, and every day through March, April, and May he marked the seedling weeds as they pushed up out of the ground. By June, 357 plants had sprouted in his weed garden.

What interested him even more than the remarkable progenerative powers of what we like to call weeds was that over those three months, more than two-thirds of the seedlings—277 weeds—were killed without human intervention. Darwin thought the weeds had simply choked each other out, until he discovered that his weed-killer was a cohort of hungry slugs.

That still leaves eighty weeds in a very small patch of ground that have to be removed by some gardener's hand. Although it would be nice to think the slugs are on my side in this battle, I suspect that they chew in service of the weeds, thinning the crop so that the plants left behind can sink their taproots deeper, spread their spiny rosettes farther, toss their seed-heads to greater effect.

I am tempted to do Darwin one better and count the weeds as I rip them out of the garden beds. Dandelions by the thousands, plantain by the hundreds, thistle by the dozens, and countless sprouts of a newcomer that I think might be Chinese lantern, though I refuse to confirm my suspicion by letting the plant go to seed.

I do what I can to keep the soil covered—grass clippings and hay in the middle of the beds where no one can see it, shredded cedar mulch around the edges because the flowers look like corsages against its dark fur. But no matter how thickly I spread it, the mulch never lasts long. By midsummer, what I laid down in spring has disappeared, leaving the soil bare to whatever is blowing in the wind.

Where does all that mulch go? Darwin wanted to know, too, so he inset a large stone in the garden, its top perfectly level with the soil surface. Slowly and steadily it sank, not perceptibly that first year, but before long, the stone was deeply embedded. By Darwin's calculation, it was sinking at a rate of 2.2 millimetres a year.

Earthworms, he concluded, were resolutely burrowing through the soil, consuming large quantities of partially decomposed organic matter, depositing their fecal casts, churning through the layers of earth, breaking them down. Based on his "worm stone," Darwin estimated that on every acre of his gardens at Down House in Bromley, England, earthworms were turning some eighteen tons of soil every year, pulling the compost down deep, and raising countless seeds into the warmth and light, where they would spring to life as weeds.

A mixed blessing. But then, what blessing isn't mixed?

There is no escaping it. If you garden, you weed.

I don't mind, really. If it weren't for the weeds, I might gaze at the garden only from a distance, like seeing loved ones from a passing train. Weeding forces me in close.

I used to attack the gardens in monthly marathons that left me discouraged and exhausted. Now I weed a bit every day, and when I can't, like a housekeeper who's let things go, I look the other way. As incentive for my daily pluckings, I keep one of those big, black plastic planting pots beside the kitchen door with my favourite Japanese gardening knife tossed inside. First thing in the morning, or last thing at night, I'll kick the bucket along a bed and dig out the dandelions and buttercups (alas!) and a weed I still haven't identified, a sapling-like growth with red stems and leaves like Solomon's-seal and no flower that I've ever seen, though its roots sink so deep I despair of ever getting them all. I keep my secateurs handy, too, because it's while I'm weeding that I notice a cherry-tree branch hanging too low over the peegee hydrangea, or a tent-caterpillar web sagging in a fork of the pagoda dogwood tree.

Don't move so fast, my mother used to say. Decades too late, I understand. I don't want to miss the scent of lilac wafting over the yard, or the drunken music of the fat yellow bees tumbling in the ajuga. The azalea that is a blur of creamy yellow from the kitchen window reveals itself, as I stroll close, to be as lovely as an orchid, its petals white and gold and frilled, like one of the Grand Girl's party dresses.

Every day my Beloved and I play cribbage over lunch on the terrace he laid the second summer we were at The Leaf. He usually wins, but I consider it a victory if I get my peg over the skunk line before he finishes the game.

"I've won a game!" I say happily, because although I've lost, I have held him to one point, instead of two.

I bring the same convoluted logic to the weeds. No matter how many are still out there, each one I've pulled today is a victory, a plant that will never set seed to filter across my beds will never rise with the worms to sprout not only now, not only next year, but like the curse in Pharaoh's dream, for seven long years.

WHAT THE FARMER KNOWS

IN MY MIND, THE GROWING SEASON is like one of those neatly coloured diagrams we used to make for science-fair projects. Spring planting → summer weeding → fall harvest. According to this neat configuration, I should be heading now from seeding into weeding, looking forward to long months of gently caring for my plants. Summer has just begun; harvest is still a glimmer in the distance, something that happens after the yellow school buses are back on our road.

If only it were like that.

In fact, harvest begins when the Farmer takes off his first cut of hay, which happens shortly after the summer solstice. Each year it catches me by surprise. I'm still seeding and, heaven knows, still weeding, when I hear the heavy scissoring of the mower in the field across the road and, suddenly, harvest-time is here.

We've been picking from the garden for months, of course. Lettuce and spinach in April; asparagus in May; snow peas and beet greens in June. These are gleanings for the table, food that will be eaten today or tomorrow. "Harvest" is something else altogether. Harvest is what the ant did and the grasshopper neglected to do. Harvest is picking for the future.

The morning that I hear the rumble of the tractor and mower on the road, I look out over the Kitchen Garden and see that the thyme is a haze of pink. The oregano has already sent up spires. The first lavender buds have popped.

"I'm late!" I moan to my Beloved as I grab my secateurs and Japanese knife. I pull an armload of baskets from the hooks above the stove.

It has been raining for weeks, but this morning, the sun is out. If the Farmer is cutting hay, a warm spell must be coming. I don't know how, but he always seems to know when the sun is about to shine or a frost is in the air. I remember waking up one night to an odd sound cutting through the silence. I looked out the window and there he was out in his field, harvesting grain, the headlights of his tractor raking our bedroom wall. Sure enough, within hours, a torrential rain was falling, beating all the other farmers' grain to the ground.

There is something mystical about the accuracy of the Farmer's predictions. I imagine him peering at squirrel scat or observing the domestic habits of wrens. Maybe he reads *The Old Farmer's Almanac*. More likely, his father or grandfather taught him to interpret the drift of the clouds, a change of scent in the air. Or maybe he's just a very good farmer with a talent for perfect timing, like a concert pianist, or a chef.

By the time I get outside, the sun has driven off the morning dew and dried up the rain. Herbs are supposed to be harvested when the fragrant oils in the leaves are most concentrated, after the plants have matured but before they bloom. I ignore the unfolding flowers and snip the French thyme, lemon thyme, and sage, piling the branches loosely in their baskets. These stems are so stiff that they hold each other at a distance, giving the leaves lots of air. Indoors, I'll set the baskets on the table by the door so I can run my hand through them as I pass, helping the leaves to dry evenly, the house filling with the same pungent smell that has fixed itself to my fingers.

I set down the secateurs, pick up the Japanese gardening knife, and bend to embrace the shrub of silver tea mint, cutting through the stems a few inches from the ground. After the mint, I go at the rest of the tea-herbs—lemon verbena, lemon balm, sweet cicely. The books tell me to bundle them up and hang them from the rafters. This looks quaint in photographs, but when I try it, the tethered leaves turn grey and rot. Instead, I spread them over a pad of newspaper on the dining-room table, where I toss them like a salad a couple of times a day. After they're dry, I'll take bowls in my lap and strip leaves from the stems as I watch *Pushing Daisies*.

I carry the baskets of herbs to the house, pausing to grab handfuls of rose petals, blue cornflower, and red bergamot, the flowers that colour my tea. Thousands of little gold-and-white daisies call to me from the chamomile hedge.

"I'll get back to you," I promise, rushing off to pick the lavender, quick, before the tight little buds open and that musky scent is lost. I tie fistfuls of the leafless flower stems with raffia and hang them on the Garden Room wall, where they'll stay until December, when I'll poke them into ribbons or strip off the purple buds to drop by the teaspoon into batches of shortbread dough.

How I would love to sit down to tea and cookies now! But the sun is out, and there is coriander to pick and freeze, parsley, too, basil to put into oil, and catnip to harvest for the cats. And that's just the beginning.

Does the Farmer ever think, Why can't somebody else cut this hay? Why can't someone feed the heifers, draw down the milk? Why does it always have to be me?

I think of the Little Red Hen, running to the pig, the cat, the rat, asking for help planting the grain of wheat she finds. She sows it, then harvests it, mills it, and makes it into a loaf of bread. "Who will help me?" she asks at every step, but no one does, so in the end, when she asks, "Who will help me eat the bread?" and the cat, the pig, and the rat all chime in, "I will!" she declares, "No, you won't. I will!"

> Does the Farmer ever think, *Why can't somebody else cut this hay?*

This makes me think of the sign we pass on our way into the city: NO FARMERS, NO FOOD.

But the Farmer doesn't seem to mind the ongoing, endless harvest. He smiles each time he passes, mowing the hay, then winnowing it into rows and baling it for the barn.

It is late afternoon. I am bringing in the last of the herbs, thinking to myself, *There, at least that's done!* when I notice the ripening berries. The man who helped us restore this stone house gave us three currant bushes as a parting gift. Every year, red berries drip off the branches like jewels off Elizabeth Taylor. A pair of robins brush close as they swoop, delirious, into the orchard and I see that the cherries are reddening, the raspberries, too.

My Beloved meets me on the path and holds out a shiny bowl heaped with dark fruit. "Look!" he exclaims. "The blackberries are ripe!"

There'll be no rest for me now. Every time I go outside to feed the chickens, collect the eggs, or do a little hoeing, I'll have to pick a bowl of berries, just to keep ahead of the birds. No time to make jelly or jam or juice. I'll dump the day's harvest into containers in the freezer with the hope that by January, maybe, we'll have something fruity to spread on our toast.

"We could buy bags of these berries, already frozen, at the grocery store," I say, eyeing the heap of blackberries that want washing and picking through.

"But they wouldn't be ours, would they?" he chides gently.

Is this why we work ourselves to a frazzle? So we can say that we eat what we grow with our own hands?

I don't need a crystal ball to see my future. After the berries will come beans, then the Chinese cabbage, tomatoes, and peppers, then the garlic will have to be pulled and ripened and the onions, then the celery, the broccoli, and the corn, more food than we can possibly eat in a year, but there are the children to consider, living in a city without gardens, and the food banks, which savour our excess. Somehow, in the middle of bringing in the harvest and putting it up or down—how is it that food is both put up and put down for the winter?—I will have to find time to keep weeding the rows and beds and seeding the fall crops of lettuce, spinach, fennel, and kale, planting next year's garlic on the same day we dig up the potatoes.

"It's all such a mess!" I say in frustration to my Beloved. "I'm planting and harvesting, all at once. I won't ever catch up."

"Sure you will," he says gamely. "I'll help."

And he will, but it won't be enough. There will always be more to do.

I look across the field to where the Farmer is climbing down from his tractor. He was at work in the fields before I came out this morning and he's still at it, long after I've quit. He waves at us enthusiastically.

"Nice evening!" he calls out.

I feel like a little girl who has just stomped her foot and, instead of being punished, has been handed something sweet.

Clearly, I have more to learn from the Farmer than when to harvest my herbs. What will I do, what will gardeners do, I wonder, when there are no longer farmers like him across the road?

MEMORY STICKS

I'M NOT GOOD WITH NAMES. The Rosarian paints pieces of lath that he prints with names and pushes into the ground under the leaves of his plants so that he always has an answer when I ask, "What is that?" I am more lackadaisical. Or egotistical. I always think I'll remember.

For years, as a result, one of my favourite summer shrubs has gone unnamed.

"I really love that plant, whatever it is," I say lamely to my Beloved every year when it blooms.

For several days after my Younger Son was born, he went unnamed, too. The blue card on his bassinet read *Baby ___*. It made me sad every time I saw Baby Blank, as if I didn't care enough to call him something. He *did* have a name—we'd chosen it months before his birth, but when we held him in our arms, it no longer seemed to fit. We had to know him, we decided, to christen him.

The plant I love has a name, too. It's just that I can't remember it. I recognize this as a failing in myself and have tried a variety of *aide-mémoires*. I've written names on bits of wood, on slices of plastic yogourt containers, on white plastic knives, on copper bands and aluminum stick-tags that I bought at great expense from a mail-order nursery. I've written in grease pencil, indelible marker, pen, pencil, and paint. Nothing lasts. The words fade; the markers disintegrate like memory.

Usually, I can find pictures to match the plants whose names I have forgotten. Every spring I walk around my gardens with my thick plant encyclopedia under my arm like Debrett's *Peerage* and reintroduce myself to my botanical friends. But this one plant, this lovely perennial that dies back to the ground then springs up to an eight-foot shrub within months, is nowhere to be found, except in my yard.

I bought it at a nursery, I remember that much. Its growth is nothing short of miraculous. The stems are hollow, like bamboo, and as thick as chair legs. It doesn't self-seed or expand aggressively. After five years, I gathered my courage and cut through the root to move a piece to the West Yard, where it thrives, too. The original is planted in the lee of the front cedar hedge, where it gets almost no direct sun. Even so, it sends up huge white plumes that last from early June through the summer, fading to a buff colour that holds its own until frost. I try to guess what it might be called. Plume Plant? Feather Flower? But there are no entries under these names.

"What is it?" I ask the Rosarian. He shrugs. His botanical interest is narrowly focused on roses, peonies, hostas, and lilacs, the flowers of the English cottage garden. My instant shrub barely gets a passing glance.

"What is it?" I ask the Frisian, who loves his plants but knows fewer by name than I.

"What is it?" I ask my Garden Guru.

"Some kind of knotweed?" she guesses.

"But it's not invasive." I had seen Japanese knotweed (*Polygonum cuspidatum*) at my Beloved's parents' house, where it formed an instant screen against the carport. "It doesn't have that kind of root."

"Look at the stalks," she says.

She's right. They are jointed in that distinctive knotweed way, but surely using one trait to identify a plant is dangerous, like saying every curly-haired kid must be related to me.

It wasn't until this spring, when I was visiting the lilac dell at the Royal Botanical Gardens in Hamilton, Ontario, that I finally found my mystery plant. The Gardens were having their annual plant sale and I spent an hour going through the pots, looking for something new and unusual.

"This is it!" I called out to my Beloved, who was eyeing the grasses.

"This is what?"

"My mystery plant. That great big shrub by the cedar hedge, the one with the fluffy white top. It's called 'Giant White Fleece Flower'! Why didn't I think of that?"

The other ladies inspecting the shelf of sale plants were staring. I was making a scene, but I didn't care. I had found it! PERSICARIA POLYMORPHA. FIVE TO SEVEN FEET TALL. SHADE TOLERANT. DROUGHT TOLERANT. CLUMP-FORMING. NON-INVASIVE. HIGHLY ORNAMENTAL. BLOOMS MAY THROUGH SEPTEMBER. Yes. Yes. Yes! As soon as I got home, I looked it up and found new reasons to love this pop-up summer shrub. It attracts butterflies but is distasteful to both rabbits and deer. Although I grow it in a dry, shady spot, it likes sun, too, and is equally happy in damp ground. It can live in even colder gardens than mine.

Persicaria is part of the buckwheat family, Polygonaceae, which makes it sound agricultural, domesticated, but this particular member of the family is such a smoke-breathing giant that one of its nicknames is white dragon.

My Garden Guru was right, as usual. One of the common names of the polygonums is knotweed or knotgrass. It is also known as bistort, tear-thumb, and mile-a-minute, but the name I like best is smartweed.

Not knotweed. Smartweed. I think I can remember that.

ON THE FLY

THE FLIES ARE HERE. That sentence should be drawn in letters ten feet high, in the undulating lines of a 1950s sci-fi horror movie poster—*THE FLIES ARE HERE!*

Blackflies are the first of the biting insects to arrive. I think of them as assault troops, unwelcome visitors from some alien land. I avoid the thought that they are born here, hatched on the slender grasses that float like Ophelia's hair in the stream. The larvae cling to pebbles under water, where they grow to adults that ride bubbles of air to the surface and fly away, straight to my naked flesh. Not *Invaders from Beyond*, then. *Creatures That Live Among Us.*

Blackflies like it cool and wet, and breed in the spring's trickling streams. By the first week of May, they are hovering around our heads, a little dopey, as if they've just got out of bed. Within a day or two they are swarming at our mouths, drinking in the carbon dioxide we exhale as we puff through the garden cleanup.

"Next year, we have to get all this done in April," I mutter through bug-splattered teeth to my Beloved. "Before the blackflies."

I say the same thing every year, and every year we are caught off guard. We know there are dozens of species of blackflies in our area. We know that half a dozen of them bite. We know, almost to the day, when they will appear. If the maple tree buds are popping, it won't be long before the blackflies are at us. So why does it always seem like such a surprise?

"Bugs bite," my city friends said, when I first moved to the country. They spoke in soothing tones, as if my fears were irrational. "How bad can it be?"

But blackflies aren't like bees, wasps, or mosquitoes. They don't inject the skin with a sharp needle that hurts for a second, then subsides. They are pool-feeders. They remove a chunk of flesh, spit it out, and sit on the edge of the little blood-lake they've made to drink their fill.

"They like fresh blood," I say to guests brave enough to visit in May.

When I first moved to my northern garden, I was determined to be tough. I had no choice: the garden had to be planted and there seemed no end to the blackfly siege. At the end of the day, I was swooning in my bed. On my shoulder, on a patch of skin the size of a loonie, I counted sixty-five bites. After that, I burned smudge pots in the garden to keep the beasts away and wore a beekeeper's net. When my Younger Son and my Elder Son got on my nerves, I'd say, "Behave yourselves, or I'll send you outside to play."

This year, the blackflies are manageable. My blood is no longer fresh and I work around them, getting up early while they are still stunned by the cool morning air. I wear bright colours, which they disdain, and slip on my net hoodie if I have to work late in the day, when their feeding reaches a frenzy. I console myself with the thought that they won't last long; their survival range is narrow. If the spring warms quickly, the black-fly season will be short.

And we always know when it is over.

"Look at all the dragonflies!" I said to my Beloved just a week ago as we sat in the Garden Room after dinner. In the angled shafts of evening light, squadrons of dragonflies were flying arabesques over the garden beds, scooping up blackflies by the thousands.

"Must be a boom year for bugs," he mused, glancing at me sideways, a wicked grin on his face. "It won't be long before the deerflies are here."

I like the endless anticipation of a garden. I await with pleasure the next arrival in the parade of tulips, poppies, brown-eyed Susans, and monkshood that announces the unfolding of the growing year. But with the same breath I curse the relentless progression of blackflies, mosquitoes, deerflies, and horseflies that march in lockstep with the plants. I want the flower show to be the best it's ever been, and the insects to disappear—which would mean, of course, an end to flowers.

I think of myself as a tolerant woman, yet my heart is apparently not large enough to embrace these six-legged fellow creatures. Like the soldiers driving out the Cherokee, like the turn-of-the-century grain farmers shooting the ravenous songbirds, I just want these bugs gone.

"Something bit me!" I say to the Frisian, rubbing at my hand. We have been cleaning out the daylily waterfall in the Woodland Garden, removing every last weed before the blooms open to spill down the slope, one of my favourite garden moments.

I think it was a deerfly, though I didn't see the perpetrator, which is unusual. Deerflies are big, not as big as horseflies, but big enough to do significant damage. When they bite, they do so slowly, landing first, meandering a bit to choose their spot, then settling in for a good, long slurp. (Like blackflies, they are pool-feeders.) There is usually time to shoo them away before they take a bite, but they keep coming back, persistent as five-year-olds.

The Frisian wears a fabric-softener sheet under his ball cap when he works. He used to stick deerfly patches on the back, but the dryer sheets work better and they're cheaper, he says. Following the latest organic lore, I spray my arms and legs with Listerine, which smells nice but has no discernible effect. I react to deerfly bites, but even so I refuse to use DEET. It melts my eyeglasses, so what would it do to my skin?

I pull off my glove. My hand puffs to twice its size. There is no visible bite-mark, but the response is typical. I once fell asleep in the Muskoka chair and woke up blinded by a bite between my eyes.

"You sure that's a deerfly bite?" the Frisian says, peering close. "There's a bad spider around here, you know. A fella in the city had his thumb eaten right off. Couldn't do nothing to stop it."

Within the hour, he has e-mailed me photographs of a thumb in various stages of disintegration, from a small scabby area to a festering sore to flesh rotted away exposing the bone. At the end is a picture of the brown recluse spider, an insect smaller than a penny that is also called the corner spider, because that's where it likes to live. (It's difficult to imagine how this makes it distinct from other spiders.) It is also called the violin spider, because of a brown patch shaped like a musical instrument on its thorax, but apparently that's next to impossible to see. And it has six eyes instead of the usual eight, if anyone is counting. Sheds, outbuildings, basements: it likes dark, undisturbed places, which hardly seems a distinguishing trait, either. This is clearly a spider that does not want to be recognized. It is native to the southern United States, but that doesn't mean it can't be in my backyard. The changing climate is inviting many insect species to travel north.

"!!!!!!!!!!!!!!" I write to the Garden Guru, when I forward the Frisian's e-mail.

She does some research: the reaction to the spider bite shown in the photographs is rare, she says, but not unknown. Then she tells me the story of her brother, who contracted a flesh-eating bacteria while clearing out a shed. He lost forty pounds during the ordeal, but managed to keep his hand.

I watch the deerflies on the screen of the Garden Room with new appreciation. They are quite pretty, really. Something that might be filigreed in gold and worn on a lapel. Why did I find their green eyes horrific? Compared to a flesh-eating virus, they are enchanting. And it's only the females that bite; the males are off collecting pollen, making my garden an even prettier place.

"What are you doing?" my Beloved says, coming up beside me to peer at the flies on the screen.

"Changing my mind," I murmur.

He looks at me oddly, as if he thinks perhaps he should be dialling 911.

"How's your hand?"

I raise it, and the flies scatter. "Just a flesh wound, sir."

When the females bite, they make a little cross-stitch incision with their mandibles, no longer than it has to be, and lap up a little blood so they can go off and breed.

I'm a mother. I can understand their need.

READING A GARDEN

"Nothing but perennials," I said grandly to the Rosarian in those early days at The Leaf. "They look after themselves."

He paused in the tour of his garden and looked at me sideways. "Is that so?" he said, cocking an eyebrow.

My experience with perennials was limited. Mostly, I had grown vegetables, which have to be seeded every year. My impression of perennials was that they came up every year, leafed out and bloomed, and died back on schedule. They were more like family than friends. Predictable. Reliable. They didn't need to be seeded and fussed over each spring. If carefully selected, they didn't produce a horde of offspring that had to be plucked from their loamy cribs.

I forgot that perennials grow. The life of an annual is obvious: an impatiens or a green bean travels from birth to death in a single season. It happens right before your eyes. Perennials follow a longer trajectory, not unlike our own, from mewling tenderness, through a vigorous prime, to a garrulous and demanding ripe age. What I saw on my tour of the Rosarian's garden were strapping plants in the fullness of their lives. I didn't know about those that had to be pried from the soil, separated from their progeny, carted off for a hasty burial in the compost bin.

I learned. The 'Autumn Joy' sedums I planted were cute cushions the first year and voluptuous, rosy mounds the second, but in the third year they grew flaccid, splaying at the centre by midsummer, stems collapsing, exhausted, to the ground.

"You have to cut out their hearts," my Garden Guru said.

I did as I was told. The following spring, I drove the transplanting spade hard into the core of the plant. I made four neat cuts, as if to quarter the sedum, stopping short of the centre and making another cut across, loosening four saucer-shaped roots. The centre I removed to the compost. No matter, I thought. Instead of one sedum, now I had four.

Perennials don't eliminate the labour of a garden so much as shift it from nurturing the young to caring for the old.

Almost a decade later, I have sixty-four 'Autumn Joy.' They're not much work: the dried stalks have to be pulled out in the spring, but the grey-green leaves never need watering. The flower heads flush pink, then rose, then a deep maroon on cue. One spring out of every four, I bend to the bed and cut out their hearts, a job as distasteful as anything these gardens force me to.

Perennials don't eliminate the labour of a garden so much as shift it from nurturing the young to caring for the old. It is a lesson the Rosarian left me to learn on my own. It's a matter of preference, I suppose, like a nurse choosing maternity over the geriatric ward. The first is filled with hope and promise. Perennials prompt darker, more disturbing thoughts.

As I bend on this lovely spring day to cut out the core of the sedums, I wonder, Does the heart always go first?

"It's like the Roman Empire," my Beloved offers. "When the strength goes to the edges, the centre can't hold." He pauses.

"Or it's like genetics," he carries on, warming to the subject of death and destruction. "Continuous reproduction from the same genetic material eventually weakens the organism. A species needs new blood to be invigorated and evolve."

Sometimes the entire plant has to go. I'm thinking of *Centaurea montana*. If I cut it back now, as soon as it blooms, it will set a second crop of feathery blue flowers, but eventually, the mother-plant will grow rangy and I'll have no choice but to dig it out. In its place, I'll keep one of the offspring that inevitably I find nesting under the leaves.

"Why is it that only the young survive?" I say, rubbing the ruddy soil from my hands.

"We'll be next," my Beloved replies brightly, forking the wilting cornflowers into the bin.

But I think: No, we won't. The Rosarian and the Humanist are ahead of us in this macabre line-dance. Behind us stretch the Carpenter; the Garden Guru; my Sister the Therapist; the Elder Son; the Younger Son and his wife; my Beloved's Daughters; the Grand Girls. How many sedums will there be at The Leaf by the time they're all gone?

I shouldn't do this, make the garden my tarot card layout, the green entrails I read to tease meaning from a life.

I can't hang on to my gloomy thoughts. They melt like Popsicles in the spring sun. I look past the sedums and my surgical secateurs to the 'Johnson's Blue' geraniums, neat mounds that are just about to bloom and will bloom again in August if I cut back the flowering stems. The delphiniums, too. These perennials demand nothing. Year by year, the plants grow stronger, the blooms taller, fuller, richer.

If I have a choice, I think, this is how I'll grow old. Not like a sedum, demanding attention, losing heart, someone else forced to take matters in hand so that I can carry on. No, I'd rather be a delph, endlessly vigorous until an act of God or nature cuts me down.

LOST AND FOUND

I LOSE THINGS MORE OFTEN THAN I FIND THEM. My keys, my wallet, my shoes on a regular basis, though I prefer to think of such temporary losses as goods gone astray that, like lost kittens, are already wending their way back to me. Not lost; misplaced.

There are some things, though, that I lose and never find again. I still think of them: the bracelet of rough-cut semi-precious stones I lost when I was ten; my great-uncle's antique carved cane, which went missing when I moved from my northern garden. Last month, I lost my favourite fountain pen somewhere in the city. This week, my secateurs.

When I garden, I carry my Japanese knife, my loop-hoe, and my red Felco secateurs. The Rosarian and my Garden Guru both carry their secateurs in leather holsters that slip over their belts. I have a holster, too, but I don't own a belt because I never buy pants with loops, which means I have no place to put my tools except on the ground, where they disappear into the weeds. I lose them and find them a dozen times a day.

"Your secateurs are probably in the compost," my Beloved suggests.

"That means I won't find them until next year!" I wail.

Last summer, we had wwoofers, young men and women who travel the world, exchanging farm and garden labour for room and board. We've had wwoofers from Japan, Australia, England, Germany, France: we couldn't manage our gardens without them. They generally come with boundless enthusiasm and limited skills. The Irish boy we hosted last July was a delight, with a keen ear for jigs and a penchant for gardening without shoes. He was eager and interested, but careless with tools. When he left, I searched in vain for my espresso coffee scoop and my best garden fork. This spring, I found them both in the compost.

Compost is an industry with gardens the size of ours. I remember walking in the Butchart Gardens on Vancouver Island and coming upon a clearing ringed with heaps of earth in various stages of decomposition. I was delirious with desire. Our own set-up is less sophisticated. At the edge of the woods, between a resuscitating Russet apple tree and the Shed Garden, my Beloved built two large, wooden five-by-five-by-five-foot bins. The sides are slatted horizontally, to encourage air circulation, and the boards across the front can be removed: when we clean out the bin, we take them off, and as the bin fills up, one by one we slide them back in place.

On the far side of the bins is an open space, a small clearing surrounded by woods. Our idea was that as one bin filled up, we would turn the compost into the second bin, and as decomposition in the second bin progressed, we'd turn it into the empty space, where it would complete its transformation to fully composted soil. Every spring we would wheelbarrow the compost to the gardens, clearing that space so the cycle could begin again.

In my mind, the system worked brilliantly, a smoothly operating machine that took in weeds at one end and delivered soil at the other. In practice, the bins fill in a couple of weeks, and the empty space, too. Whenever we get a minute, we clear off the top foot or so of slimy vegetative matter from one of the bins, scoop out a barrow or two of good compost, then carry on heaping up the weeds.

Maybe the bins are too nestled by the woods. The Himalayan impatiens and Jerusalem artichokes block what little light there is, so that the mass stays frozen well into May and doesn't really start to cook until July.

"This year, I'm going to get the compost under control," I vowed to my Beloved.

I reduced the vegetable plantings to make room for a large compost heap in the Winter Garden. He set up metal stakes that he wrapped in chicken wire. The space was the size of a child's bedroom.

"We'll never fill that," he said.

I thought he might be right, but here it is, early summer, the harvest has scarcely begun, and already the pile is heaped so high with wild parsnip, lambsquarters, and rotting dandelions that if I stood in it, I'd be up to my neck in slithery weeds.

Are my lovely red secateurs in there, too?

This week, the Frisian has been emptying the bins and wheeling barrow after barrow of dark, crumbly soil to the garden beds, where I spread it as a top dressing around the base of the perennials and shrubs. I have found a dishcloth; two plastic-coated scouring pads, one blue, one yellow; a tea strainer; and three forks, not the sterling but the silver-plated ones with the rose on the handle that we inherited from my Beloved's mother. Utensils often get swept along with parings into the compost bucket—potato peelers, small knives, the occasional spoon—but when the Frisian hits shards of china and odd bits of corroded metal, I know we've gone past the compost into topsoil that has been secreting things for decades.

In the course of ten years of digging at The Leaf, we've found enough broken china to glue together a service for eight, enough bits of harness fittings to bridle a four-horse team. I rarely push in a spade without unearthing something: fencing wire, rusted pliers, a plastic airplane, bottle caps, once an unbroken glass inkwell, twisted canning jar lids, old liniment bottles still brown with unnamed medicine. The Forge Garden got its name from a cache of more than a hundred square nails we found buried in a heap within a thin scrim of soil over a large flat rock where, we thought, a forge might have stood. We've found enough bricks for half a dozen paths, and in the gentle slope that fronts the property, square-cut blocks of limestone that my Beloved heaved behind the house and set into the fine, low curve that defines the Terrace Garden.

We tell ourselves stories about each thing we find: The boy who pitched his plastic warplane into the air and searched for days among his mother's brown-eyed Susans, never finding the downed sniper. The woman who dug a hole at the edge of the woods and buried her garbage every week, broken canning jars and Noxema bottles when she was young, and later, when her knuckles were swollen and sore, the shards of cups that slipped from her fingers to shatter on the floor. The man who learned the hard way how to make a decent nail with a hot fire, a hammer, and an iron rod, pitching his rejects into the scrub where they were buried in the drift of centuries.

The Rosarian and the Humanist have found arrowheads in their gardens. Their property ends at a stream, where a flat limestone alvar would have made a perfect camping spot for the people who had this place to themselves before Europeans claimed it. I imagine a Native family pitching their teepees there for the summer, hunting deer in the forest of oaks that must have grown in this area once, maybe fishing for trout in the stream that perhaps was a river then, tossing aside the chipped, triangulated stones that weren't quite sharp enough or precisely the right shape.

My diggings have never unearthed much of monetary value, except in the city garden, where I found a penny that was minted long before this northern lakeshore became a nation with a name on the map of the world. Most of what my hand-tools clang against is worthless, except as a marker of who might have dug in these beds before I came along. In the jar lids and plastic toys and hand-forged nails I see the history of my gardens: I like to think that the earrings I've lost there, even my secateurs, are serving the same purpose—laying down a history of my time in this place for some future gardener to excavate.

A dear friend, an older woman, once gave me a silver ring etched with the primitive shapes of animals, like the cave paintings of horses and bison on the walls of Lascaux. I loved that ring. I never took it off. One evening, after a day of transplanting, of whipping off my garden gloves to loosen the roots with my fingers, settle them in the ground, I discovered the ring was gone. I was distraught. My friend had recently died of cancer, and I felt I had betrayed her with my carelessness. I rented a metal detector and scanned that small city garden. The detector gave off a cacophony of beeps as it moved over the water pipes, the gas lines, all the technological detritus buried in the ground. I tossed the useless thing aside and moved methodically, plant to plant, feeling among the roots.

I found it, eventually, but as I slipped it on, I thought I should have left it where it lay, something lost, waiting in a garden to be found.

TOUCH-ME-NOT

"Plants can move," I tell the Grand Girls when they come for their summer visit.

"No, they can't," they say in unison. Oh, for the conviction of a child!

I don't tell them about *The Power of Movement in Plants*, Darwin's last book and a subject that fascinated him to his death. Instead, I guide their fingertips to the pointy-ended swellings that hide under a canopy of blooms. They scarcely touch it and the capsule explodes, pitching seed into the air as the pod coils back onto itself.

"Touch-me-not, it's called. Or jewelweed."

"Why?" they say.

"For the leaves, I think. Look how fresh and green they are, and they stay that way all summer." We scrape at the leaf with our garden-ragged fingernails. I explain that the upper surface is coated with a water-repellent that keeps the leaves shiny. "That's called a 'cuticula.' And see how silvery they are underneath? That's from thousands of tiny air bubbles trapped in the plant flesh."

"Like ginger ale," says the younger of the two.

"Like your hair," says the older one, remembering the time

we looked at our skin and nails and hair under a bug magnifier, the strands of their hair solid with rich chestnut and a golden red, while my white filaments were clear and bubbly as champagne.

We are bent over the Banquette Garden, the bed that surrounds the McIntosh tree in the middle of the West Yard. Chartreuse hostas and creeping astilbe punctuate the circle, with impatiens in the spaces between.

"Busy Lizzies," I say. "That's their other name."

Most years I collect seed and start my own impatiens seedlings, but last year, for reasons I don't understand, there were few seed pods and so I picked up flats of Busy Lizzies from the prison greenhouse, red and white and a deep salmon pink.

"But what's their *real* name?" the older girl insists. Of the two, I suspect she will be the gardener, though the younger may well love flowers more.

"*Impatiens walleriana*. Because they can't wait for the pod to dry and split and drop the seed. They're so impatient to reproduce that their pods burst open at the slightest touch."

I don't tell them that I grow impatiens because I am impatient for their bright burst of colour. Or that I buy them from the prison greenhouse because who knows more about impatience than men confined behind razor-wire?

I used to feel a little embarrassed planting impatiens, even the somewhat exotic New Zealand impatiens that I bought one year. Impatiens are like petunias or geraniums—showy but ordinary. Mundane. Proof, I used to think, of a certain lack of imagination and will in a gardener.

But there are almost a thousand species of impatiens. Aside from the low-growing, mounding touch-me-nots that are a staple of local nurseries, there are the balsam impatiens, a tropical branch of the genus that looks nothing like its North American cousin.

"What's that gorgeous plant?" I whispered to my Beloved one afternoon at our favourite restaurant in the city. We were sitting on the outdoor patio, enjoying a Scotch (him) and decaf Americano (me). We often do this: ask a question that the other can't possibly answer.

While he was trying to figure out which plant I meant, I was breaking open sugar packets and emptying them onto my saucer. My Beloved knew what I was up to; he studied the sky as I headed for the flowers that bloomed on tall, fleshy stalks. I slipped a seed pod into an empty packet and squeezed. Pop. Another packet, another seed pod. White sugar packets for white flowers; Sweet'n Low for the pink. Pop. Pop. Pop.

"Do you know the name of that plant?" I asked the waiter. It was obviously an impatiens—I could tell by the touch-me-not pods—but what kind?

"No idea," he said, "but please don't pick the flowers."

"I wouldn't dream of it," I said.

Now I grow balsam impatiens (*Impatiens balsamina*) in the Pine Garden, in the Shed Garden, and between the elm stumps.

"These aren't Busy Lizzies!" the Grand Girls insist when I show them.

It's true: they look nothing like the impatiens the inmates grow. Instead of low mounds of bright colour, the plants stand three feet tall. The leaves are toothed and spiral around the stem; rose-like blossoms cluster at the leaf nodes in tropical shades of fuchsia and white. I think they must be a jungle bloom, for they hardly seem to need any light. I grow them in the perpetual shade of the Arbour Garden, behind the perennial foxglove.

"Sure they are," I insist right back. "You're a girl," I say, pointing to one and then the other, "and you're a girl, too. One is tall and one is short, but you're both girls, right? So these are both impatiens, even though they don't look anything alike."

I point to the seven-foot hedge of tiny, nodding pink flowers that screen the compost heap, too. The stems are hollow, as thick as bamboo; the leaves are vaguely heart-shaped and big as dinner plates. "And that's an impatiens, too."

I got the seed from friends who share their apple-cider press with us every autumn. We take chicken-feed bags bulging with windfalls to their house and, in the course of a brisk late-summer afternoon, press a couple of hundred litres of cider that we split between us.

"What are those gorgeous flowers?" I asked on my way back from the compost, where I'd dumped yet another barrow-load of apple mush.

My friend shrugged. "I don't know. Take some if you like. But they self-seed like crazy. That's why we keep them by the compost."

Which is where mine grow, too. The huge seed pods of the Himalayan impatiens (*Impatiens grandiflora*) propel fistfuls of seed in all directions, and every one of them, it seems, successfully sprouts. I spend hours each spring ripping them out of the Shed Garden, with some finding a home as far away as the Arbour and the Winter gardens.

"They're called Himalayan impatiens," I tell the girls.

"Because of the mountains," says the older girl, who likes to spend time with the atlas.

"Because they're tall as mountains!" says the little one, beaming.

"They're Jack-in-the-Beanstalk flowers," says my Beloved, joining us from the path in the woods that leads to his writing cabin. "Jack didn't get beans in exchange for his mother's cow. He got touch-me-not seeds."

The girls wrinkle up their noses at this new piece of information. They're not sure what to believe, then they decide it doesn't matter. They laugh.

"Tell us the story!" they say.

BOUNTY

LIVES OF TREES

AT THE LEAF, TWO MAGNIFICENT ELMS cool the Croquet Lawn and cast the Arbour Garden in deep shade. Their trunks rise fifty feet to thick, horizontal branches that reach gracefully in all directions. A person could build a house under one of those trees and never be out in full sun.

"You better plant some understorey," my Garden Guru said the first time she came to visit. "Those elms won't last long."

But they did. We cut a path through the Arbour Garden close to the two great elms so we could walk by them when we needed something from the shed. The bark on an elm is thickly furrowed, like an old elephant's skin; touching it, you have no doubt this rising pillar is a living thing. Because elm roots lie close under the soil, I was careful what I planted near the base: a cluster of low ferns, some rose-balsam for summer colour, a fringe of sweet woodruff and primrose for spring. On the edge, a sweep of creeping Jenny to hold back the grass until I figured out what else to do.

The University of Guelph in southwestern Ontario keeps a registry of elms that appear to be surviving Dutch elm disease, which the scientists refer to by the chilling acronym DED. The elms die from a fungus spread by the elm-bark beetle. The fungus itself is native to Asia: it acquired its name when Marie Beatrice Schol-Schwarz, a Dutch phytopathologist, identified it in 1918. Ten years later, it was discovered in the United States, inadvertently imported on elm burls for the furniture industry. By 1967, it had arrived in Ontario.

Many trees have a strong enough immune system that they can live for twenty or thirty years before succumbing to the fungus, so it isn't unusual to see young elms. But occasionally a tree is truly resistant, living to sixty, ninety, over a hundred years, with a trunk so wide that it would take three people to span it, hand to hand.

The problem is, these big old elms are too scattered and too isolated to make a difference to the gene pool, so researchers have developed a kind of dating service for lonely elms. They collect cuttings from mature, resistant elms and graft them to produce new trees that are then planted in a seed-producing orchard where they can breed with their own kind, producing, it is hoped, a highly resistant strain of elm.

We have lots of young elms along the fence row, at the edge of the woods, and dotting the Arbour Garden. The two at the far end are clearly of a different generation. They soar above the canopy, their branches reaching across the Croquet Lawn to shade half the Forge Garden.

"The Elm Recovery Project is looking for survivor trees," my Garden Guru says. "When the trunks get to be seven feet around, we can register these two."

And so every year my Beloved and I wrap the measuring tape around the stately elms. This summer, one of them is less than an inch shy of seven feet. The other is even closer.

"Next spring," my Beloved says with confidence. "They'll be big enough then."

Weeks later, the leaves begin to yellow. Flagging, it's called. By September, the bark is falling off in great, curving scabs.

The Woodcutter arrived this week. He cuts and splits and stacks our winter firewood in the far field, about twelve face cords, culled from the woods behind the house, which are a mix of Carolinian and St. Lawrence forest—maple, oak, ash, ironwood, basswood, and a stand of butternut we didn't know we had until the Woodcutter pointed it out. Fuelwood is supposed to be cut by March, before the spring sun prompts the sap to rise. That way the wood is good and dry by fall.

"I been awful busy," he tells me, shaking his head when he climbs down from his truck in early August. He is a big man, with a body built on bacon and eggs, and eyes the colour of a blue jay's wing. "Thirty year back, we lost all the elms to the Dutch disease. Then the young ones sprung up and now they're dying—bam!—all at once. I can't keep ahead of 'em."

The Woodcutter gets his chainsaw out of the back of his truck. He takes out a file and sharpens the teeth, balancing the machine on his knee. A flick of his hand and the chainsaw is roaring, biting into the dead wood.

It takes less than ten minutes to bring the first one down. The ground shakes; branches break with a sound like shattering glass. By noon the trees are cut, split, and stacked in pieces with the rest of the firewood. I'm supposed to haul the brush to the side, but I pause and lean over the stump to count the rings, losing track somewhere around a hundred. A century undone in a morning.

We did plant the understorey trees, as my Garden Guru suggested. I am, after all, a belt-suspenders-and-Velcro kind of woman. I added a serviceberry tree, a hop tree, a couple of pagoda dogwoods. Last fall, when it was clear the elms were flagging, I put in a spindle tree, a huckleberry, and a chestnut oak.

I can't seem to plant enough trees. After a summer when the temperature stayed above thirty degrees Celsius for six weeks straight, I bought a couple of Carolinian species in anticipation of global warming: a Kentucky coffeetree, an Ohio buckeye. I am desperate for an American beech, which according to my Garden Guru plays host to more species of insects and birds than any other. To celebrate our first anniversary at The Leaf, I bought my Beloved a purple birch, thinking this would become a tradition. I planted a mountain ash when the first Grand Girl was born in September, loving the idea that the berries would ripen every year for her birthday, and a winter-blooming witch hazel when the second arrived one February. All three have succumbed to insects, disease, or my ignorant neglect. Only the trees I planted for my parents—a ruddy Norway maple for my father and a weeping willow for my mother—continue to thrive.

"It's dangerous to make a symbol of a tree," says my Beloved. "Better to plant something we'll live to see."

The trees I added to the Arbour Garden are still slender wisps. They look like skinny ten-year-olds beside the strapping young-adult elms.

"Won't be long," the Woodcutter says, pulling a rollie out of a flip-top plastic cigarette case and bending to a match. When he straightens, he nods his head toward the sky above the Arbour Garden. I look up and see that the leaves on a branch of a young elm are a sickly yellow. "You'll lose them, too."

I suppose he's right. He works with trees all day long. Like most people born and raised in the country, he does a bit of everything: grows soybeans, drives a truck, makes maple syrup. Mostly, he fells trees.

When he arrived at seven this morning, I told him I'd been up since five, weeding the gardens. "I been up watching the deer and the turkeys play," he said. "That's what I do every day, get up four, four-thirty, right at the break of day, and stand on my deck. That's when the deer come out, and the wild turkeys, with their little ones."

He's a tender-hearted man, too tender-hearted to harvest tree corpses for a living.

"I'm glad now that I saved that ash," I say, trying to find some good in the wreckage of the elms. Years ago, when we were clearing out the Arbour Garden, I had my hand on the stem of a sapling growing within a foot of one of the elms. I was about to rip it out, give the lovely elms room to breathe, when my Beloved said, "Why not leave it?" and so I did.

It's a good-sized tree now, so big that every spring I wonder if I shouldn't try to move it, in case it is crowding out the elm, leaching its water supply. Then I remember about the emerald ash borer, the beetle imported in a packing crate to Toronto that is now making its way through the ash forests of the province. My little tree probably isn't worth moving: its days are numbered, too.

"I suppose the ash borer will get it," I say to the Woodcutter, imagining our few forested acres bristling with skeletons.

The Woodcutter stubs his cigarette and flashes a sharp smile.

"That's no ash; that's a butternut," he declares, then his voice softens. "That tree'll be around for a good long time yet."

PLAYING CHICKEN

THE CHICKENS HAVE NO NAMES. This is not a matter of principle: I can't tell them apart. They are uniformly brunette, a rich reddish brown inherited from their Rhode Island Red ancestors. They are plump, too, the legacy of the other branch of their genetic tree, the solidly rounded Barred Rock. Sweet-tempered, with a fondness for back rubs, they provide us with eggs and manure and do yeoman's work clearing the gardens of earwigs, grubs, and slugs.

The hens arrived at The Leaf as day-olds peeping inside a cardboard container the size of a cake box. Their training began immediately. I picked each one up in my hand and whispered, "Here, chick-chick," in what I hoped was a mother-hen voice. For the first month they lived in the front porch, inside a wire pen bedded with straw that I replaced every other day. Chickens are born knowing how to eat and drink and flop in a heap on their bellies, wings spread, like a mass of pale yellow puppies. "Here, chick-chick," I cooed as I filled their feeder twice day.

My Beloved and I moved them into the henhouse when they were fully feathered, or when we couldn't stand the infernal cheeping and the stink any longer, I don't remember which. Every day I would dip a small tin pail into the bag of chick starter mash and rattle it in front of me as I backed across the chicken yard. "Here, chick-chick-chick," I'd cluck, louder now, and they'd come running like a pack of wobbling, short-winged velociraptors. As they pecked at the seed, I'd rub their backs, working my way to the base of the tail. They'd squat and spread their wings, raising their beaks toward the sky with something that resembled bliss.

"They don't want a back rub; they want to be trod," my Beloved observes dryly. If he smoked a pipe, he would be taking it out of his mouth at this point and tapping the bowl significantly against the gatepost of the chicken yard. Chickens are stupid, he says. "It wasn't the California condor people were thinking of when they coined the phrase 'bird brain.'"

I forgive him. He doesn't stand in the dusk and listen to the hens on their roosts, cooing their evensong. He hasn't watched them go still as reeds when a hawk strafes the lawn, their beaks pointing skyward for so long even I have trouble seeing them in the shrubbery or against the grooved bark of a tree. When they find a hole in the fence and scatter across the yard, he tries to herd them back to the coop.

"You can't herd chickens," I say. "They're too smart. They want to be seduced."

We've never had a cockerel—the friends who take our eggs are undone by the red spots in the whites—so my Beloved has never seen for himself the complex dynamic of the poultry harem. Thirty years ago, in another life, working in my northern garden, I heard a cry of triumph and ran with the hens to where the rooster had a milk snake by the tail. As I watched, the snake flexed its body and sank its teeth into the red flesh of the rooster's wattle. The hens rushed to his defence, darting at the snake, pecking, until it let go its death grip. With a toss of its head, the rooster flipped the snake in the air, opened its beak, and swallowed the creature whole. For an hour, the cockerel strutted about, treading one hen after another, the snake writhing visibly in his crop.

Here at The Leaf, I let the laying hens run free in the spring, a tonic after their long winter in the coop. They clean the gardens of insects, harrowing the ground with their powerful claws. I appreciate their digging when I'm preparing the beds, but once I seed and plant and the gardens are mulched, I resent the mess they make. I keep them locked in their compound, throwing greens and worms and small snakes over the fence. In the evening, I let them out for a walkabout, calling them to a particular patch of ground that I don't mind being disturbed, tidying up after them as they trail back to the coop to roost.

The hens often complain about the arrangement, running along the fence and bleating to be let out, but Amelia is the only one who ever flatly refused to stay put. I'd be off among the roses, digging in bone meal and flaxseed, or pulling dandelions out of the daisies, when suddenly there'd be a rush of wings, a thump on my back, and there was Amelia, rocking on my shoulder. She'd hop on and off, depending on what I was uncovering in the ground at my feet. When I moved from bed to bed, she'd cling to my shirt like a plump red parrot. At the end of the day, I'd put away my tools, pick up Amelia, and take her back to the coop to spend the night with the others, who shuffled along the roost, reluctantly making room for her, keeping themselves distant, as if they didn't trust a bird that spent so little time with its own kind.

My Beloved bought higher poles, tacked on another five feet of chicken wire, but still Amelia flapped and soared. There was no fencing that chicken in.

We have turkeys at The Leaf, too—wild turkeys that make forays through our yard, a herd of about two dozen or so. *Flock* is the correct term, but that collective seems too small, too tame somehow, more suited to chickadees and chickens than to creatures that stand as high as my chin, their feathers flashing bronze in the sun. There is something prehistoric about the turkeys as they march down through the Woodland Garden and fan out across the yard, aiming for the raspberries and blackberries, whatever fruit or seed they can find.

"Dinosaurs," my Beloved says. He tells me again the story of Phil Currie, the Canadian paleontologist, who cleaned the bones from a Thanksgiving turkey and developed the theory, now widely accepted, that dinosaurs still walk among us in the form of birds.

The wild turkeys act like they own the place, which I suppose they do in a way. *Meleagris gallopavo* has lived in North America for about ten million years, except for a short hiatus during the twentieth century, when overhunting and deforestation extirpated them from this part of the world. The birds have since been reintroduced and they seem happy to be home. My hens, on the other hand, are descended from *Gallus gallus,* the red junglefowl native to the tropics. They don't belong here at all.

On an afternoon in early August, the turkeys emerged from the woods just as I was going inside to sharpen my secateurs. The sun was about to set. Amelia was on the Croquet Lawn. The turkeys were between her and the henhouse, where the rest of the layers were having their dust baths and the young meat birds were trying out their adolescent voices.

Amelia was edging toward the forest, bending nonchalantly to peck now and then. She had obviously seen the turkeys. Her plan, it seemed, was to circle around behind these gargantuan feathered creatures as they moved toward the house. With luck, she'd be safe in the henhouse before they realized she existed.

Just then, an old tom spied her. The thin black beard on his chest quivered. He lifted his head and fixed one eye in her direction. Another turkey did the same, then another, until all those arching necks were ramrod stiff, heads cocked toward Amelia.

She stopped.

They advanced.

Turkeys can't seem to move their legs without jerking the muscle that attaches to their heads. As they closed the circle around her, they punched the air with their blood-starved beaks. Amelia stood stock-still, as if hoping they might not see her plump red form silhouetted against the spring-green grass.

I was watching from the back stoop. The old tom took a step forward. Leaned his head down. Retracted his neck like a boa before it strikes.

"Amelia!" I shouted. "Here, chick-chick! *Here!*"

I thumped my chest with my fists and reached out my arms. The turkeys turned to look, and in that split-second of inattention, there was a frantic whirr of feathers, a panicked flapping, and Amelia came flying across the yard, propelling herself into my embrace, clasping at my chest.

The turkeys, in their stately way, turned and fled.

I carried Amelia to the henhouse and set her down on the straw. She squatted at my feet, wings akimbo. I rubbed her back hard, working my fingers deep into the feathers. When I stopped, she gave herself a shake and flew up to the roost, losing herself among the others, just another chicken.

THE SCENT OF A GARDEN

WHEN WE FIRST MOVED TO OUR HOUSE in the city, I had little to do with my neighbours. Our household was big and busy, that was my excuse. But it seemed necessary for survival, too, to barricade myself from the eight cats next door that took my garden as their litter box; from the family who brawled under the steady gaze of the Virgin and two of Snow White's dwarfs tucked in their niche in the stone wall.

The grotto at the back of our yard became my refuge. There, the walls of an old limestone quarry rose high on two sides and a tall cedar fence blocked the third, though it shut out the one neighbour I liked, the one I called to as we hung out our laundry in the soft grey pearl of early morning.

"I have something for you," my neighbour said one day, hoisting a plastic grocery bag. "A plantain lily," she said, as if she could hardly believe she'd remembered the name. Her husband was the gardener in the family. He had died suddenly not long before.

I knew almost nothing of flowers. I had cultivated a vegetable garden since I was twelve, but I'd rarely grown flowers and had never planted anything ornamental whose primary beauty was its leaves. I took the bag from my neighbour and, later that day, shoved the clumps of rhizomes in the ground by the arbour my Beloved was building to shield the entrance to our private little grotto.

"They'll grow anywhere," my neighbour said. "That's what my husband used to say."

Within a month, thick shoots were rising out of the ground, unfurling leaves the size of elephant ears, so deeply ridged they'd hold a rain for days. I'd never seen anything like it, except perhaps in the *bosques* of Brazil, where I grew up, my childhood years giving me a taste for the overblown, the exotic.

By midsummer, three-foot scapes were rising, branched and budding, from the knee-high mound of leaves. When the pure white, five-inch trumpets burst open late one August afternoon, I fell to my knees, breathing in a fragrance that intensified with the fading light, drawing down an aromatic miasma that transported our little grotto out of the clamouring city to some moist and mysterious island where Odysseus would surely have lingered, had he been caught by that scent.

As I grow older, I choose my plants as much for how they smell as for their colour and shape. I moved to that city in June, and when I'd bicycle the streets, I'd be knocked off my seat by the scent of lilacs and peonies leaping over fences, bounding into the street, refusing to be contained. The whole town smelled of flowers. The bedrock around there is limestone, and lilacs love limestone. They grow wild in the fields, whole hundred-acre farms gone over to lilacs, so that in late spring, the country roads are like tunnels through the hedges and those of us passing through inhale what a bee must inhale, its whole body vibrating with pleasure as it rolls among the blooms.

When we moved to The Leaf, I brought a node of the plantain lily with me. By then I knew it wasn't a plantain, and it certainly wasn't a lily. It seemed to be a hosta, though it bloomed long after the others and the flowers weren't mauve but white, larger and more intensely fragrant than any hosta I'd ever met.

Scent, of course, is a purely practical matter for a plant, a reproductive strategy, like colour and the arrangement of its petals. In fact, it is an odd axiom of the garden that the prettier the scent, the less impressive the bloom. One or the other will do, if it's bees and beetles a plant is out to seduce. The deliciously scented dame's rocket (*Hesperis matronalis*), for instance, isn't much to look at, and the fragrant lemon lily (*Hemerocallis lilioasphodelus*) has the smallest, narrowest trumpet of any species I grow. 'King Alfred' daffodils may be pretty in a vase, but it is the button-bloomed narcissi that can perfume an entire room. Sweet-scented stocks, white nicotianas: for all their lovely fragrance, they don't make much of a show. Which is why I was so enamoured of my neighbour's plantain lily: it was fragrant and lovely, too.

The plantain lily grew, but slowly. My new gardens were large and sweeping; I could hardly wait for those drifts of white trumpets, a glowing hedge that would light up the August night. At an end-of-season plant sale at a local nursery, I found a dozen plants that looked just like my August lily. PLANTAIN LILY, the label said. I bought them all, caught in a visionary tsunami of fragrance moving nightly across our yard. I rushed home and planted in a frenzy of excitement. The next year, barely a whiff rose from their paltry blooms. The flowers were white and somewhat larger than the usual hosta offering, but nothing at all like the five-inch trumpets that open on my neighbour's August lily to blare scent across the late-summer garden.

They couldn't both be plantain lilies. Or could they?

When Engelbert Kaempfer, a botanist with the East India Company, described the first hostas three hundred years ago, he named them according to what they looked like: *vulgo gibbooshi Gladiolus Plantagenis folio,* which translates as "the common hosta with the plantain-like leaves." A century later, the first hostas were sent to Europe as a gift of seeds from the French consul in Macao to the Jardin des Plantes in Paris. Soon the hostas were growing by the thousands in the public gardens in France. Botanists originally put it in the genus *Hemerocallis,* along with the daylilies, but the name was eventually changed to *Hosta plantaginea.* Gardeners, breathing in its penetrating perfume, called it "Funkia," a name that faded with its popularity.

After its Victorian heyday, it vanished from gardens until Gertrude Jekyll rescued it, setting the 'Grandiflora' form—a sterile hybrid introduced from Japan in 1841, shortly after our house was built—in large Italian pots, mixed with white lilies, hydrangeas, and ferns.

Since then, a kind of hosta fever has gradually taken over domestic gardens. Hundreds, maybe thousands of varieties have been developed: hostas with blue leaves and yellow leaves, striped and mottled leaves, purple flowers as well as white. Life is full of trade-offs, and plant-breeders, it seems, opt for colour and size over aromatics. Miss Isabella Preston's lilacs (*Syringa* x *prestoniae*) don't sucker as much; they bloom later and are hardier, but the flowers scarcely smell at all. Modern roses, sweet peas, lilies—so many lack a clear scent that one wonders if the gene for fragrance isn't attached somehow to those that govern susceptibility to black spot and rust. Or maybe it's just that we no longer need pretty-smelling flowers to strew on our floors and hold to our noses. People don't have much natural smell to them anymore, either.

There is something personal about this business of smell. What I find lovely, my Beloved abhors. The mint that hangs to dry in the kitchen, for instance, smells fresh to me, but has him sniffing the air for incontinent cats. Jasmine I always thought exuded an irreproachably lovely fragrance, though Gertrude Jekyll tells the story of her brother writing from Jamaica about a large-flowered variety: "It does not do to bring it indoors here, the scent is too strong. One day I thought there was a dead rat under the floor and behold it was a glass of fresh white Jasmine." Jekyll herself fostered an almost pathological dislike of the smell of barberry: "It was so odious that it inspired me with a sort of fear, and when I forgot that the Barberries were near and walked into the smell without expecting it, I used to run away as fast as I could in a kind of terror."

I have a friend, the Poet, who suffers from anosmia: she has no sense of smell at all. I know this about her, though I keep forgetting. The loss of smell isn't something that exerts itself into a relationship, the way a loss of sight or hearing might. The other day I took her a stem of August lily.

"I can't smell a thing," the Poet said.

I pushed the trumpet up to her nose. "Are you sure? Nothing at all?"

She shrugged and shook her head with regret. "But it's beautiful," she said.

When I got back into my car, the lily's sweet perfume still lingered in the air, and that evening, when I walked back from closing in the chickens and paused by the Kitchen Garden to watch the fireflies, I picked out the scent of each of my night-bloomers—nicotiana and four o'clocks, the last primroses and the first blooms of *Datura* 'Belle Blanche,' and above it all, through it all, the heady perfume of my neighbour's August lily.

What a shame to erase such scent from the gardener's bouquet! Wasn't the hosta already fine enough as it came to us from the meadows of southern China? It is, in fact, the only hosta from China, the most southerly of the wild species; it alone among the hostas loves sun. It blooms late in the season, a month or more after the other hostas, and it keeps setting new leaves all through the summer, which means it never looks completely riddled by the incursions of slugs. It is also the only hosta that opens its blooms at night, which is why it was once called the curfew lily.

But the breeders couldn't resist. Those pale, ribbed leaves were just too plain. They fiddled in their laboratories to come up with multi-bloomed, striped-leaved, giant and miniature plantagineas, until now there are over fifty registered hybrids, at least two dozen of them common to the nursery shelves. 'Royal Standard,' the one I bought at the end-of-season sale, was the first to be patented. Later I wanted to rip them out of the garden, but I stayed my hand. It's not their fault they are pale imitations of their parents. Instead, I transplanted them far from the house, in a bare patch by the Emerald cedars, where they look pretty enough and I hardly notice that they give off such a paltry scent. I never think of dividing them, or giving them away.

But my neighbour's August lily, my lovely old *Hosta plantaginea* 'Grandiflora'—this is a plant I cultivate with care and aggressively share. It came to me the way plants have always come to gardens: a node, a rhizome, a corm, a seed, a cutting, a seedling handed friend to neighbour to friend.

"Our gardens told each other everything," Colette wrote. The French writer grew up in her mother's gardens at Saint-Sauveur-en-Puisaye, where plants and vines rose over the stone walls to cast their fragrance into neighbouring yards.

I think of this as I divide my August lily, setting some in my own soil, saving a few to hand on to friends, wondering what secrets they will tell, and to whom. A garden, I've come to realize, is never entirely one's own.

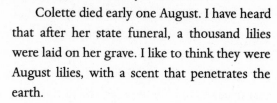

A garden, I've come to realize, is never entirely one's own.

Colette died early one August. I have heard that after her state funeral, a thousand lilies were laid on her grave. I like to think they were August lilies, with a scent that penetrates the earth.

HOLLYHOCKS

You may choose whom you spend your day with, who sits down to share your meal, who comes to your bed at night, but there is no choosing love. It erupts wild, in inappropriate places, where it must know it can never grow, though it does, defiantly, without care for whether it is needed, or useful, or desired, until it is part of your landscape, whether you like it or not.

I look at the hollyhocks and think of that kind of love, unbidden, unwelcome, with a stubborn will all its own. Love for an unkind parent, a wayward child, a betraying friend. Not sexual love but a love that carves deeper channels with its tough, seeking root. The kind of love I think of when I see the hollyhocks break through the earth.

I have tried to weed them out. They don't grow well here. They start out lush enough—rosettes of fleshy leaves that might, in other places, be sauteed and scattered over rice. I get at them with a trowel, a shovel, a long transplanting spade, digging out the gnarl of root, saying, like a surgeon, *There, I got it all.* Yet all the while, back behind the spirea or under the spreading yew, a smaller rosette is taking root, hiding itself from my fierce and watchful eye until one morning there it is, seducing me with its blossoms, tissue petals the palest peach, the sunniest soft yellow, sometimes a red that leans so close to black it might be blood.

Who can yank out a flower that opens itself like that, a slow unfolding to the centre, the dark and knobby core? Not me. Not even when I know I should. When I can see, as clear as day, that the little Colorado spruce will suffer in the hollyhock's pushy shadow. When it sits there like a stupid cat, right in the middle of the path.

The Rosarian welcomes the hollyhocks, which have spread to a swath against the high cedar fence that encloses one wall of his rose garden. They rise up and over: you can see them from a distance, waving spires above the soft nubs of Bourbons and Damasks, a counterpoint, a welcome part of the composition, not central but essential, like a maiden aunt with stories to tell.

Perhaps the seed came from him, blown on the wind. The breezes that wash over my garden wash first over his. But not always. Not today. Today the rain is lashing from the east, the path our worst storms take, unexpected and predictable.

Maybe the hollyhocks were planted by Apple Annie. I must remember to inquire. Or maybe it was her mother who pushed the first seeds into the ground there under the library window, thinking what a pleasure it would be to glance through the glass while going about her work and see the thick thrust of green stem, barnacled with bloom.

One spring I gave in. I'm not quite sure when it happened— some years ago now, though not so many that I should have lost track. *Have it your way,* I said, throwing in the trowel.

I made adjustments. Planted a trio of 'Goldflame' spirea under the library window, between the stone wall of the house and the spread of low cedars. Now, by midsummer, the shrubs are tall enough to hide the hollyhock's unsightly leaves, thinned and pitted with the rust borne by the fungus *Puccinia mal-vacearum*. The flowers are all I see, thrusting up like memory. Sometimes I pick them for a bouquet, but mostly I leave them be. They are like a friend I no longer trust but can't stop loving, and I am careful where I let them in.

Still, I think how much worse off I'd be now if I'd succeeded in ripping out every last root and bud, if these beds were filled only with flowers of my choosing. Nothing wild. Nothing to come upon unexpectedly. Nothing that tears *me* up by the root.

The hollyhock may not be among my favourites, but I need its rough reminder: love, even when it lasts, is never planted; it comes unbidden, borne on the wind.

GYPSIES

I saw my first Gypsies in the South of France, in Saintes-Maries-de-la-Mer to be exact, a small town in the Camargue, that vast grassy delta at the mouth of the Rhône. Saintes-Maries-de-la-Mer is named for the three Marys who found Christ's tomb, and who left Judea with their uncle, Joseph of Arimathea, landing on the Gallic coast, some say with the child of Mary Magdalene and Christ. Twice a year, Gypsies congregate in the small French town to honour their patron saint, Black Sarah, the Egyptian servant of the Marys. We arrived in Saintes-Maries-de-la-Mer on the first of these two holy days and followed the Gypsies as they lifted Black Sarah's statue from the church and carried it above their heads into the Mediterranean, paying homage to the place where she first set foot. We watched as they sang and danced pinwheels in the square.

What does this have to do with onions? Well, *Gypsy* is short for Egyptian, and Egypt was once considered the ancestral home of the Roma. The race is now thought to be of Hindu origin, but wherever they came from, Gypsies have always been a people on the move. And so I think of them, their swirling dances, their plaintive songs, their lavish costumes, whenever I pass my Egyptian onions, the exotic, restless allium that walks across my garden, refusing to stay put.

Allium is one of the largest plant genera in the world. There are more than a thousand species, including the wild leeks that we dig from the woods in mid-May and the lovely Persian *Allium aflatunense* that sets blooms like purple lollipops among my desert yucca in late spring. But the staple allium in my garden is the common onion, both *Allium cepa,* the bulbing onions that are drying right now in the shed before we braid them for the winter, and *Allium fistulosum,* the bunching onion that spices our table for three-quarters of the year.

Bunching onions, spring onions, green onions, scallions: these are all different names for the plant most commonly known as the Welsh onion. Their leaves rise in hollow tubes—*fistulosum* means hollow—from bulbs that look more like a slightly thickened, blanched stem than an actual bulb. Some gardeners harvest small, immature *Allium cepa* onions and call them spring onions or green onions, but these lack the defining characteristic of a true Welsh onion. Only the Welsh cluster in bunches.

I always assumed that Welsh onions were indigenous to Wales, the land of leeks and song, but no, "Welsh" derives from an old English word, *welisc,* which means foreign. The English called the onion "Welsh" because it came from China.

The Welsh onions at The Leaf come from a clump I started from seed in my city garden, years ago. I dug up a few scallions to bring with me when we moved, planting them first at the edge of the Rockery, then in the Winter Garden, then in the Kitchen Garden so they'd be closer to the house. Last year, when I remodelled the Kitchen Garden, adding my tea and savoury herbs, I repositioned the Welsh onions so they would flank the entrance, where I can get to them even before the path itself is clear of snow.

From early spring until the ground freezes solid, I harvest the slender shafts, loosening the clump with a garden fork, extracting what I need, and settling the rest back into the soil, deep enough that a considerable length of the shaft is blanched. So long as I leave one in the ground, there will be a fresh clump of twenty or more the next year. I grow 'Evergreen White Bunching,' a variety true to its name. It scarcely dies back at all, even in our cold climate, so it is always the first thing I harvest from the garden come spring. I chop it into salads and stir-fries, the white "onion" and the leaves, too, even the oniony-sweet blossoms. Like asparagus, *Allium fistulosum* is one of very few perennial vegetables. Unlike asparagus, which we gorge ourselves on for a few weeks then dream about for eleven months, the lovely Welsh onion supplies our table through three seasons of the year.

The same is true of the Egyptian onions, except that I didn't plant them. These gypsies appeared one year out of nowhere, or so it seemed—an odd onion with a distinct underground bulb like a small cooking onion and the hollow leaves and bunching habit of the Welsh. In fact, they are a cross between *Allium cepa* and the Welsh, and it was my lackadaisical harvesting habits that introduced them to my garden. In August, when I pull the globe onions for winter storage, I inevitably miss a few. They overwinter under the mulch, and when they sprout the following summer, I leave them be, figuring survival should be its own reward. The rogue *Allium cepa* is biennial and sets flowers its second year, at the same time the Welsh onions are sending up their white ping-pong-ball blooms. The bees do the rest.

The Egyptians are stronger-tasting than the Welsh, not as delicate in salads or laid beside the cheese, bread, and pickle in my Beloved's ploughman's lunch. But I don't grow them to eat, except in the very early spring, when they are tender and mild. No, I keep the gypsy onions because I love their wandering habit, their August acrobatics.

Walking onions, they're sometimes called. Or topset onions. Or tree onions. As they mature, an odd growth appears at the top of the stem, encased in a shroud that peels back like a snakeskin, revealing a clutch of small, dark-skinned bulblets that send out thin green scapes to dance on the breeze like a many-armed Indian goddess. As the bulblets grow, the weight of them bends the stem toward the soil. Where it touches, the little bulbs sprout, making a new garden for themselves, slowly but purposefully moving away from where I planted them, wandering in search of something better, or maybe just different.

The Welsh onion, though delicious, seems staid by comparison, stolid in its unfailing upward thrust. Or maybe it's me who lacks imagination when it comes to vegetables. It never occurred to me to blow into one of its hollow leaves, making a Welsh-onion flute like the one I heard played on YouTube, in an orchestra of instruments that included a carrot violin, a cucumber trumpet, and a radish slide-whistle.

Perhaps when I'm off in some other part of the garden, the Welsh onion holds its leaves to the breeze and makes music for the gypsy onion, twirling in the sun.

GARDEN TRIALS

CONSTRAINED IN ITS LITTLE BLACK POT, the mint looked delectable. And it *was* delectable: I sniffed at the leaves, then pinched one off and nibbled. Sharp but sweet. Like spearmint, but without the bitter aftertaste. Delicious.

I wanted more.

It was the compactness of the little square-stemmed beauty that seduced me. I tucked it into the Pine Garden, not far from the kitchen door. All summer, I snipped leaves for lemonade, and in the fall, I added a handful to my harvest tea so I could savour that hint of sweetness through the winter, too.

The second spring it emerged in its usual place, still a neat bouquet.

"What a well-behaved mint!" I enthused to my Beloved one morning at the breakfast table. I pointed discreetly to his chin.

"Well behaved, is it?" he replied, brushing crumbs from his beard. "Your highest compliment."

I do like a plant with good manners. Take the basket-of-gold alyssum, for instance. It takes up a little more ground every year, but tentatively, visibly, as if waiting for permission. It never assumes it has the run of the garden, as sundrops do, or lily-of-the-valley, or creeping Jenny.

I don't banish unruly plants, but I'm strict with them. Every fall I rein them back, slicing my spade deep through their eager, thrusting roots. The self-seeders are harder to manage. They toss their seed profligately, forcing me to my knees to pinch out their offspring by the thousands. I loved them at first, these spreaders and seeders: they filled up my new beds with instant greenery. I've kept a few favourites—those whose bullying ways are balanced by a gorgeous scent or long-lasting blooms—but mostly I got rid of them, replacing them with perennials that keep to their place.

Why, then, buy another mint? It wasn't as if I were a mint virgin. Just a year before, I'd spent an entire day digging every last piece of silver mint from the old Tea Garden. I'd grown tired of unlacing its thick white roots from around the angelica, the lemon balm, and the rue. In their new home in the Kitchen Garden, the mints are confined to big, bottomless pots sunk deep into the soil. "Let them get out of that!" I said to my Beloved as I tamped the earth triumphantly around the rims.

Yet here I was in year two, admiring my new mint, fully believing it was different.

This is not the first time I've lost my mind. I planted *Impatiens grandiflora* in the Woodland Garden. I actually bought a red Norway maple and installed it in the Forge Garden, where its limbs would extend over wood-chip paths, the perfect medium for its million maple-key babies. I planted mother-of-thyme (*Thymus serpyllum*) next to elfin thyme (*Thymus praecox*), thinking they'd get along, as if mothers and daughters ever do. I introduced sweet woodruff among the ferns. I opened a well-tilled bed in front of the *Rosa rugosa,* a hedge with irrepressible wanderlust.

I know better. I know the invasive tendencies of that mother of all thymes, the urge to procreate that has made a scourge of the Norway maple. I know the character of these species and yet I plant them anyway, as if they are paper cut-outs from the garden design book my Beloved gave me one Christmas. I spent a week playing with the colourful cut-outs, slipping them into slots in a cardboard garden bed, spires of lupins at the back, daisies in the middle, sweet mounds of geraniums in front.

"Nice garden," my Beloved said, innocently.

"Hah!" the Garden Guru snorted when she saw it.

We both knew the lupins would sprout among the daisies, the daisies would die out at the centre, and the geraniums would spread like a bad cold, my careful design undone.

"If only it would stay," I said wistfully, the same thing I say every time I clean the house, or get a good haircut, or start a perfectly congenial friendship.

Hair grows, dirt accumulates, friendships wax and wane. There's an inevitability to gardens, too. They shift and change. Maybe that's what makes me think the character of a plant is pliant. That it will behave differently *this* time, in *this* garden.

I wonder: What are these plants like when they're at home? Lily-of-the-valley (*Convallaria majali*), for instance. It has been around so long that one of its common names is Our-Lady's-tears, from the legend that the flowers formed as Eve left the biblical garden, weeping. It is native to northern Europe, but apparently there is an indigenous colony in the eastern United States, too. If lily-of-the-valley behaved in those woods as it does in my garden, nothing else would grow there. Something, some intense pressure in its home soil, has made it aggressive. Beautiful but pushy. What happened, I wonder, to all these invaders to cause them to develop such bold, relentless means of insisting on survival?

This spring, when I raked the fallen cones and branches off the Pine Garden and dug out the dandelions, loosening the soil around the coral bells, the astilbe, and the lady's-mantle, I found a thick white root coiling through the earth. The odour was unmistakable.

"You can't change a leopard's spots," my Beloved mused as he hauled away the wheelbarrow full of errant root. I asked him to bag the mint so it would melt in the heat, but he's throwing it on the compost. "It'll be fine," he says, as he always does.

All spring and summer I battled with the mint. It's almost fall and still I find bits of it pushing up out of the soil with all the eager bravado of a May sprout. I curse myself, for I have brought this on myself.

"A mint is a mint is a mint," I chant under my breath, as if that will forestall my gardener's amnesia, force me to remember that a plant is what it is, and always will be, no matter how much I want it to be something else.

I'm not alone in this affliction, this penchant for forgetting the true nature of living things. When I visit the Garden Guru, we tour the front-yard garden she made to replace the lawn that came with the house. Sweet woodruff runs rampant along the foundation. Perennial sweet peas spring up everywhere.

"You planted those?" I ask, astonished.

"It was an experiment," she replies. That's what she tells her students when a planting fails. But I want her to come clean with me.

I press further. "But you must have known they would spread and reseed like crazy."

She shrugs as she bends to the woodruff's heady fragrance. "I guess," she says dreamily, clearly caught in its spell. I always thought she was immune to the impulses that plague me as a gardener, but then she repeats the words that succour me, too. "But it's worth a try."

HIDE AND SEEK

THE MOMENT WE SAW THE LEAF, we knew it would be home. The rolling fields, the woods at our back, the stream at our feet, the stately stone house, the meandering road. Two hundred years ago, this part of the province was surveyed and parcelled out to refugees from the American Revolutionary War. According to the original plan, our hundred-acre plot was defined by county roads on its northern and southern limits, but the stony landscape resisted. Instead, the settlers continued to use the old Native carrying road, even though it cut their farms in two. Today, the farmhouses still hold to the high ground to the north, across the road from their slope-footed barns. Our barn was torn down before we arrived, but the house stands where it always has, just a hop and a skip from the passing parade.

"Too close to the road," my Beloved said. He'd imagined a long, shady path to a house hidden in the trees.

"We'll plant a hedge," I said brightly.

Plant cedars started off close to the top of all we planned to do that first year, but it was still nothing but a promise when the snow started to fall in December.

"It's too late," my Beloved said glumly. "Now we'll have to look at the traffic for another year."

Traffic on the road, we have discovered, consists of the handful of people who work in the village, an occasional tractor, the milk truck, and the school bus. But I took his point. I spent an afternoon locating a hedge-planter who agreed that it really wasn't too late. A few days later, he showed up in a blizzard with a cultivator and a trailer full of cedars. By nightfall, a thin green line staked a barrier between our house and the road.

The hedge grew, better in the sun than under the spreading maple tree. The chunks of buried limestone, leftovers from a centuries-old fence row, didn't seem to bother the roots at all. The planter had churned up a trench a hundred feet long and at least two feet wide; I did my best to keep it free of weeds. In the spring I heaped it with compost, and in the fall I heaped it even higher with leaves.

"Time to prune the hedge," my Garden Guru said the second summer.

I protested. "But I can still see through the branches!"

"Pruning encourages growth," she said, the way my mother would say, *A penny saved is a penny earned,* or *A stitch in time saves nine.*

She showed me the terminal buds on the tips of the plants. "Sun-loving plants grow where there's the most sun, which is at the tips. If you take off the terminal buds, the plant will generate new buds all along the stem."

My Beloved flinches each time I get out the shears. I can tell by the way he squints at me that he doesn't believe me when I say, "I have to cut it so it will grow."

"No one cuts the cedars in the woods and they seem to grow okay," he says.

I ignore him and put my faith in the Garden Guru. First thing in the spring, I give the sides of the young hedge a light trim, just a half inch to nip off the outer buds and encourage the inner ones to grow.

"It may slow the plants down a little now," I explain, "but they'll catch up by summer."

Late in the spring, I get the Frisian to even off the tops. When we moved in, there was a short length of hedge across one side of the house. The trick is to keep the older cedars in check until the youngsters fill out and up, especially the ones lagging in the shade of the maple.

In the heat of summer, cedars go dormant. They put what energy they have into firming up the new growth they put on in the spring. It is now, in early August, before they send out their second flush of growth in the fall, that I give them a major pruning, trimming the top and sides, cutting out dead wood and misshapen branches, but leaving the old wood at the centre of the bush. I work to restore the nice cone shape—a narrow top to shed snow and open more of the branches to the sun—then I give them a light sheering to stimulate buds that will still have time to harden off before winter sets in. If I can get to only one pruning a year, this is it.

We planted other hedges, too. A thirty-foot row of lilacs—started the first spring from suckers I removed from around the white, mauve, and deep-purple French lilacs near the house—divides the east lawn from the road, and a hedge of Arctic willow lines the driveway. The lilacs I leave to their own devices, but the willows I take down each spring to within inches of their gnarly trunks.

"They'll never come back," my Beloved moans.

"Yes, they will," says the Frisian, who likes the shears almost as much as I do.

"Have faith!" I say, but my Beloved averts his eyes every

time he leaves the house until the first thin willow leaves appear.

There is a short hedge in front of where we park the cars, too. For the first few years, I ignored it. *Go free,* I thought. But then it grew spindly at the bottom, so I asked the Frisian to lower it, a three-year project that involved lopping off three feet at a time, until now it is full again and a sensible height.

Sometimes, crouched over the cedars with my shears sharp in my hand, I'm tempted to whimsy. Why not a spiral? Or a pyramid. Maybe a rabbit for the Grand Girls. Or books: how about a hedge shaped like a library shelf?

But I am not Edward Scissorhands. I have no gift for topiary, and though I like the notion of shaping a shrub, cedar pyramids and yew rabbits seem about as far from nature as a gardener can get.

Or are they? Is a trio of shrubs pruned to the shape of running hounds any less "natural" than a swath of rhododendrons on an Ontario lawn? Or for that matter, how natural is it to build high green walls around a meadow just because there is something I don't want to see on the other side?

In Carol Shields's *Larry's Party,* an ordinary man discovers a unique talent for creating labyrinths or mazes. The two are defined quite differently: a maze is a puzzle with branching paths, while a labyrinth has a single, circuitous route to the centre. Shields plays with the idea: though life seems like a maze, it's really a labyrinth that leads inexorably, though not without confusion, to the heart of the self.

But why hedges? Why a *living* maze?

There is a cedar maze near The Leaf. For years, I drove past it without realizing what it was. I thought it was a hedge, until one day I looked back from a height in the road and saw that in fact, the hedge had three dimensions: its walls stretched back into the field and enclosed other walls, walls within walls that seemed, from the roadside, a solid block of shrubbery.

I wish I had gone in. I have never meandered through a maze, though I can imagine the green quiet there. It would be a puzzle, finding my way, but a kind of pilgrimage, too. A place that challenges a person to control her own fate. This way or that? Or maybe stay, breathe in the resinous air. Losing track of direction isn't always an undesirable thing. It's what I love about walking in the woods—not knowing exactly where I am going, or where I am.

The house beside the maze is for sale now. The hedge is overgrown, its top no longer crew-cut, its walls erratic and bushy. I wonder if I could still push my way through. It would be like walking in a jungle. Like walking blind. Even our widely spaced hedges do this for me: they block out the world. Make me stop and look close.

OF NECESSITY

WHEN THE NIP IN THE AIR BECOMES TOO MUCH for the flies, when the trees ignite and the monks lift their indigo hoods, it is time to kill the chickens.

I'm not squeamish about the killing. When I was younger, in an earlier life, I killed dozens of chickens in a day, chopping off their heads in the maple grove, dipping them in a steaming iron kettle, plucking, gutting, then cooling them in a bathtub full of ice water that swam with pale, puckered flesh. I knew where my Sunday roast chicken came from. But having done it, I don't want to do it anymore. The work is hard and it's distasteful. I've had enough.

"Let's find someone to do it for us," I say to my Beloved as we gaze out over the flock.

"Let's do it ourselves," my Beloved says. It is his first year raising chickens. "We should finish what we've started."

He finds a stump the right height and drives in two nails. I sharpen the axe. He sets up the barbecue to keep the water at the boil. I find the canning kettle and hope it is big enough. He sets up the folding picnic table. I gather the plastic cloth and the cutting board, then sharpen the gutting knife. I fill the kitchen sink with water and dump in trays of ice.

We start early, but it's hard to get up before a chicken. They are already milling about the yard, cooing and clucking—a dozen layers and two dozen pullets that we've raised from day-olds for meat. They are all the same breed: dual-purpose Rhode Island Red crosses. Their feathers shine in the morning light like a teenager's freshly dyed hair.

"Which one first?"

"Whichever one is easiest to catch," I say, resigned to the day ahead. They'll all end up in the freezer. What does it matter?

My Beloved and I work together to corner the first one. She knows something is up. The rest of the chickens are flapping and cackling, trying to distract us. But once he grabs her and tucks her under his arm, she goes quiet. Chickens hate being stalked, but they don't mind being caught.

He strokes her back as we walk together across the meadow to the chopping block.

"Lay her head between the nails," I say.

"She'll get away."

"Trust me, she won't even try." I don't like knowing this much about how a chicken dies.

He puts his hand under her throat and positions her head between the spikes, then lays the length of her out on the stump. She doesn't resist, only rolls her eyes to fix him in her gaze. He grips her feet with one hand, as I suggest, and with the other swings the axe above his head, bringing it down hard in a clean cut across her neck. She jumps to her feet and runs into the grass.

"Oh, I forgot to tell you about that," I say.

My Beloved is looking aghast at the headless chicken flailing in the meadow. I run after her and pick her up by the feet. She flaps her wings a few times, then her body goes still. I start toward the barbecue.

"I'll pluck if you don't mind the killing."

"Let's do it together," he says, leaning the axe against the stump.

We dip the bird in boiling water until the feathers let loose, but not so long that the flesh starts to simmer, then we lay her on the table, each of us at a wing, and pull the flight feathers out by the handful. It's a relief to be rid of the feathers. Plucked, the hen looks almost like something from the cooler at the grocery store.

"You want to gut?" I ask. I'm not sure how much of this he is up for.

"You go ahead," he says. "I'll watch."

I pick up the gutting knife I received as a gift from a butcher thirty years ago, after spending a very pleasant afternoon in the back of his shop, watching him carve our pig into chops and roasts. He wielded the knife like a scalpel, paying homage to the animal with his skill. When it was over, he gave me the knife, a thin, long blade with a rubber handle that fits my grip exactly.

He wielded the knife like a scalpel, paying homage to the animal with his skill.

The knife is sharp. I start slowly, taking off the wing-tips, the feet. I cut the oil sac from the base of the tail, then slide the knife down beside the neck vertebrae to remove the crop and release the gullet and the windpipe. Finally, I turn the bird belly-up and make a long incision from breastbone to cloaca, the opening that serves for both egg-laying and elimination of feces. Just before the cloaca, I make a sharp right and carefully carve a line around the anal aperture. I slide my hand into the abdominal cavity, just under the ribs, loosening the guts gently from the skeletal structure. At the far end, I crook my fingers around the viscera and pull hard. There is a sucking sound and the guts give way, dumping neatly into the garbage pail under the lip of the table. This is textbook gutting: no nicked intestine, no leaking gastric juices, no foul odours.

"It doesn't always go this well," I say to my Beloved, feeling vaguely shocked at how easily it has all come back to me after ten years away from this kind of slaughter. I pry the lungs, liver, and heart from inside the cavity and inspect them. The lungs are pink, the liver a smooth maroon, the heart a taut muscle. I put them in a bowl for the cats, then hold the bird up for my Beloved, torch at the ready, to sear off the last of the body hair.

"You want to catch the next one? " I say as I carry the bird into the house to chill. "I won't be long."

By the time I come out of the house, he is walking across the yard with the next bird hanging headless from his hand. We settle into a rhythm: catch, kill, pluck, clean. We don't say much. The chickens, after the momentary squall when another of their number disappears, return to browsing, cooing, and clucking.

"Remind me, why are we doing this?" I ask halfway through the afternoon.

"Because we're carnivores," he says grimly. The air is thick with feathers. "Because we like to know where our food comes from."

I can hardly stand to think of eating this flesh. Still, I know that these chickens, raised on grass and fresh air, will taste a thousand times better than anything we could buy. Recently, we tried to purchase a six-pound roasting chicken, and in store after store, clerks told us they no longer carry chickens over three pounds. Colonel Sanders won the war before any of us even knew there was a battle.

The chickens we are killing are three months old. I keep my grandmother's old kitchen scale by the sink, and as I bag each one, I mark the weight on the outside of the plastic with a grease pencil. Six, six and a half, seven, occasionally eight, and once, nine. We have twenty-four pullets to process. We'll give some to our children, we'll trade some to friends for wild salmon they bring in from the Stikine, and we'll keep a few on hand as gifts. Some I freeze whole for roasting. The rest I cut into legs, breasts, wings, and backs that I bag in portions for the two of us.

There is a smell to this that gets into the nostrils and lives there for days. A funk of wet feathers, chicken poop, seared flesh. "I never want to eat chicken again as long as I live," I say to my Beloved as the sun lowers in the sky.

These are young pullets, their wattles just beginning to turn red. They haven't yet begun to lay, though when I gut them, I find long strings of yolks like graduated pearls set loosely in a necklace, ranging in size from tiny teardrops to the one closest to the cloaca, already encased in albumen and a soft, translucent shell. I think of the indisputably triumphant cackle with which a hen announces she's laid an egg, the way she settles back into the straw after her boasting outburst, nudging the egg underneath her, pecking at my hand when I try to steal it away.

"Let's call it a day," my Beloved says. I don't argue.

The last clutch of hens gets a reprieve. My Beloved has to go to the city, and I don't feel like killing on my own. He spends the day in meetings, where one after another, his colleagues lean in close and ask with some concern about my well-being. Finally, a friend takes him aside.

"Your glasses," he says, pointing to the lenses. "They're splattered with blood."

When he returns, we finish the job. We're keeping half a dozen of the pullets as layers, so it doesn't take long. While I finish gutting the last bird, my Beloved opens the gate to the chicken yard and releases the survivors to peck in the gardens, a treat after all they've witnessed.

But they don't head for the lettuce or the tomatoes. They leave the late-bearing raspberries alone. Within minutes, they are at my feet, pecking bits of dead tissue off the grass, nibbling at the guts of the hens they snuggled up to on the roost last night. Fiercely protective of the ovoid life they produce, these birds are indifferent to death. Their own or the death of others, it hardly matters.

I find this chilling, and oddly appealing, too. To be utterly unconcerned with how things turn out must be a relief. It would change everything. No more working like frantic ants to store up food for the winter. No more pruning in hopes of better harvest. No more transplanting here, there, everywhere, to create some short-lived effect. For days I think about this as I lean against the fence, watching the survivor-layers saunter and peck about the yard, both of us, the hens and me, locked thoughtless in the present.

FUNGUS AMONG US

ONE MORNING IN THE FIRST SPRING we lived at The Leaf, we awoke to the sight of a stout woman bending into the grassy edge where the lawn met the woods. A basket rested on the ground beside her.

I sauntered over, wondering what, exactly, one says to a kind-looking, white-haired woman poking in the grass at the back of one's yard at 6 a.m. Was she deranged? Wandered off, perhaps, from some nearby facility? Or just lost?

"Good morning!" I said cheerily.

"Oh, good morning," she replied, straightening.

She seemed harmless enough. She was neatly dressed, no food stains down her blouse or grass stains on her pants from spending the night out of doors. I eyed her basket.

"Morels!" I exclaimed.

"Why, yes," she said proudly. "Aren't they lovely?"

I was flummoxed. Morels are prized by French chefs above all mushrooms. They're called miracles in Kentucky because they once saved a family from starvation. Morels aren't just treats; they're treasures. To take a morel without permission is not thievery; it's a travesty.

"It's a shame," she went on. "There aren't as many morels as usual this spring."

"Do you come here often?" I stammered, sounding like a bad movie.

"Every year," she said brightly.

"We live here," I said, at a loss as to how to proceed.

I don't know what showed on my face, but she suddenly stopped and clutched her basket to her chest. In the end, she hurried back to her car without apology, taking her full basket of morels with her. She never returned.

Since then, every spring, my Beloved and I scour the woods where we saw her picking, but we find nothing. Perhaps she was a sprite and I am being punished for my ungenerous thoughts. For a year or two, a few morels appeared at the base of the rotting stumps of apple trees downed by the Great Ice Storm of '98, but never enough to fill a basket.

We often come upon mushrooms in the woods—puffballs, boletes, hedgehogs, shaggy manes. I used to hunt mushrooms in the woods behind my northern garden and know enough not to eat an *Amanita muscaria* or an inkcap if drinking alcohol, but my Beloved is unconvinced of my skills.

"People die every year from eating mushrooms they pick in the woods," he says archly. "Let's not add our names to the list."

He has no qualms about eating other unearned harvests. In May, we pilfer the colonies of wild leeks (*Allium tricoccum*) that sprout in the woods. There is no mistaking ramps, as they are sometimes called: they are the first big splash of green across the forest floor. The bulbs look like little green onions and taste like onions, too, with a hint of garlic. "If it doesn't smell like onions, it isn't wild leek," the Garden Guru told me the first time we went looking.

My Beloved and I came of age reading Euell Gibbons, the man who said that gardeners throw away the most nutritious crop when they pull the weedy purslane and amaranth from among their spinach plants. I used to take *Stalking the Wild Asparagus* into the woods behind my northern garden, imagining "wild parties" like the ones Euell and Freda would throw, the guests dining on frog's legs, stinging nettle, and Elder Blow fritters.

Euell Gibbons was a wild-plant prophet to back-to-the-landers, whose parents thought they were crazy—though their grandparents, raised like Gibbons during the Depression, saw the wisdom in their scavenging. Then food became sophisticated and wildness disappeared from the menu, but it's back again, and Gibbons is, too, in deluxe editions.

Wild foraging has become so popular that some places have declared wild leeks "a species of conservation concern." In Quebec, just two hours' drive from The Leaf, the government has declared it illegal to sell, import, or serve wild leeks in restaurants. People like us are allowed to harvest fifty bulbs for personal use. I recently heard a radio guest compare eating a nice-sized wild leek to dining on an old-growth cedar. I suppose letting city folks into the woods to strip the ramps is a bad idea, but we apply the same rules to our wild allium as we do to the Welsh onions in the garden: never pull more than a few bulbs from a cluster, and make sure to leave the one that raises the stem that produces the seed.

We could harvest fiddleheads, too. Plucking fiddleheads doesn't endanger the plant in the same way, since people seem to know enough to take only a few of the fronds each plant sends up. Three of the fiddlehead ferns grow here—ostrich fern (*Matteuccia struthiopteris*), cinnamon fern (*Osmunda cinnamomea*), and royal fern (*Osmunda regalis*), but they are picky to prepare.

I'd have to boil them twice, changing the water in between, then cook them thoroughly. They are twice as high as blueberries on the antioxidant scale, I know, but all that goodness comes at the price of too much work.

Jerusalem artichokes grow wild in the meadow, but I avoid them, too. They give me gas. Nevertheless, every September my Beloved digs a few handfuls and throws them into a stew along with juniper berries he brings back in his pocket. If the Farmer's Son has been lucky with his crossbow, the stew will be venison.

"Why not mushrooms, then?" I say. "Everything else in there is wild."

But he draws the line at wild mushrooms.

Early one September, while meandering through a city market, I happened upon a mushroom seller.

"'Shrooms! 'Shrooms!" he barked into the crowd, but no one was stopping.

Arranged on the counter of his makeshift sidewalk stall were several white stumps encased in plastic. I peered through the condensation.

"What are they?" I asked.

"Mushroom stumps," he said proudly. "Just open the top a bit and before you know it, you'll be harvesting mushrooms— shitake or oyster, your choice."

Eating mushrooms almost seems like dining on the earth itself: the flavour they impart is musky, addictive. They are fraught with toxic lore, I know, but they are rich in selenium, potassium, and three of the B vitamins—riboflavin, niacin, and pantothenic acid. I had a Dutch landlord once who invited my young family for dinner and served us fried mushrooms with bacon on toast. It became my favourite lunch.

My mother used to pour them from a can and sauté them for stroganoff (one can mushrooms, one can cream-of-mushroom soup). She was a woman of her time and I am a woman of mine. I buy them fresh at the grocery store, but even so, I ache for mushrooms that I don't have to scoop from a cardboard box.

I bought the stump, which turned out to be a compressed column of wood chips infused with spawn. For two weeks it sat on our kitchen counter, then the mushrooms started to sprout, lovely sweet shitakes, handfuls of them, pounds of them. We had mushroom omelettes and fried mushrooms on toast, mushrooms with every vegetable for weeks. They weren't wild, they didn't taste of the woods, but like the chicken, corn, broccoli, and onions on our plates, we had harvested the mushrooms with our own hands, and there was pleasure in that.

It is near the end of September now. The stump harvest is over, but outside in the woods, the mushrooms are still going strong. The final weeks of summer have been so wet that every morning I find fairy rings on the Croquet Lawn. My Beloved finds me crouching over them, staring at them with longing.

"If you're sure," he says gently. "Maybe just one."

OF ICEBERGS AND ITALIANS

AT HOME, I KNEW WE WERE HAVING a fancy dinner when my mother cut the iceberg lettuce in wedges and dropped a dollop of Thousand Islands dressing on top. The rest of the time, we ate our iceberg broken into a bowl with Kraft French dressing that was life-preserver orange. We never ate "greens."

Iceberg lettuce was the vegetable I most desired as a teenager. Minus the dressing, it was one of those negative-impact foods, like celery and gum. Chewing burned more calories than it contained, so just by eating it, I was guaranteed to lose weight. The poor iceberg has fallen from grace, replaced in the produce aisle by romaine, Boston, leaf lettuce, mesclun, and any number of exotic greens, but I admit that I miss it—its watery crunch, its pale heart.

I love lettuce. It is the lemonade of vegetables; nothing else satisfies in quite the same way. And few plants in the garden balance utility with beauty so perfectly. Lettuce (*Lactuca sativa*, named for the milky juice it exudes when cut) is a huge family, but most of the varieties gardeners grow fall into three main groups: *Lactuca sativa* var. *capitata*, or head lettuce, which subdivides into crispheads like iceberg and butterheads or looseheads such as Bibb and Boston; *Lactuca sativa* var. *longifolia*, which are the long-leaved romaines; and *Lactuca sativa* var. *crispa,* which are the looseleaf lettuces that form open rosettes but not actual heads.

Around the first day of spring, as soon as the snow melts and the ground is thawed, I sprinkle the first seeds. My favourites for early spring are 'Reine des Glaces' and 'Merveille des Quatre Saisons,' Batavian lettuces with rather thick leaves that grow in loose rosettes. You'll never see them in a grocery store because they don't ship well, but even among gardeners in North America, the Batavians are not well known. That's a shame because these crispheads are both relatively heat-tolerant and can stand the cold. One year—an unusual year, I admit—we ate our first 'Reine des Glaces' on March 24 and I picked frozen leaves from the garden on December 25, thawing them to crunchy perfection for our Christmas dinner.

Batavians, to my taste, offer the best combination of all types of lettuces. Technically, they are crispheads like iceberg, so they are crunchy with rather thick ribs. But the leaves are relatively long, like romaine; they're soft like a looseleaf; and at the centre of the rosette, they're almost as buttery as a butterhead. Best of all, the Batavians have real flavour, something that can't be said for many lettuces. Guests sometimes ask if I've used a nut oil in the dressing when I serve 'Reine des Glaces,' it has such a distinctive taste.

'Ice Queen' is a green Batavian. I hold a slight preference for 'Merveille des Quatre Saisons,' partly for its exuberant name, but mostly for the colour of the leaves, a crinkly bronze at the tips, deep rose near the centre, almost chartreuse at the base. This is a very old French heirloom variety, with a big floppy rosette and a heart that, though small, is solid and sweet. It doesn't stand the heat too well, but it doesn't mind the frost, so it is the one I choose to carry me through these darkening days of fall.

The bright yellow-green of 'Black-Seeded Simpson' is the perfect complement to the bronzy red 'Four Seasons.' This is an old standby, introduced in 1850 and a staple in seed catalogues ever since. It matures more quickly than the Batavians, though I tend to harvest all my lettuces as if they were loose-leafs. I sow the seed thickly by tossing it in a band along the edges of the garden paths, then I thin out the leaves for salads (sometimes in great fistfuls), leaving a few plants, well spaced, to form heads.

The deeply red, low-growing 'Lollo Rosso' is also a lettuce-bowl and garden favourite of mine. It has tightly curling, crinkly leaves, less than half the size of a 'Simpson' leaf, though if the surface area were measured, it would be close. I love the furry look of it bordering the Kitchen Garden path: once I alternated red and green 'Lollo Rosso,' but the effect was overly Italianate for my taste. The green adds little new to the salad bowl in terms of colour, texture, or taste, and can't compete on the health front: the red contains the antioxidant quercetin, believed to reduce the risk of heart disease.

By mid-July, my spring-planted lettuce is bolting. I don't mind. I pull it up and feed it to the chickens, who seem to like its sun-warmed bitterness. I leave in place the strongest plant of each type and let it go to seed for next year's harvest. As soon as the seeds form, I pull the plant up by the roots and hang it upside down inside a paper bag; otherwise the seeds will be lost to the soil, where they'll be killed by the cold. I've learned the hard way that it is wise to label each plant, because once they go to seed, they all look the same. If there is a lesson in this, I don't want to know it.

In Italy, a few years ago, I came across a monk planting lettuce and basil seedlings into small raised beds set among some tall pines.

"Why?" I asked, raising my eyebrows, shrugging my shoulders, and lifting my hands, palms skyward in the universal gesture for *Please explain.*

He pointed up to the waving branches, down to his fragile seedlings, and put his hand to his heart. Because they love the shade.

Now I plant a few heads of lettuce in the Pine Garden, too. Together with other greens—mizuna, arugula, kale, Chinese cabbage—they get us through the dog days of summer, but it's late September now and my August sowing has sprouted. I lift a half dozen plants of each variety and set them in the rectangle of the small cold-frame my Beloved built. It's still too warm, but soon we'll glass them in, then bank the frame with straw, and finally, as winter approaches, drape the glass with blankets at night. I can almost taste the buttery 'Black-Seeded Simpson,' the 'Lollo Rosso' that rises red as the setting sun, and my 'Marvel of Four Seasons,' which, with luck, will see me through to Christmas.

THANKS GIVING

IF MY GARDEN WERE A NOVEL, the climax would build now, as the trees turn to flame against a thrilling autumn sky. All the plotlines are converging: Will there be enough potatoes for the winter? Will we get the last beans in before the frost? Who will come home to celebrate the harvest?

All through the growing season, I feel a hand at my back, nudging me forward, but now it is a definite push. The story is drawing to a close, the last pages will soon be turned. I am reluctant to leave this narrative. I want to linger through its final moments, but the pace has picked up. We're heading pell-mell to the end.

The Grand Girls load pumpkins into their child-sized barrow and wheel them one by one to the kitchen, where their mother scoops out the seeds and roasts the flesh for pie. Their father, my Younger Son, cleans the seeds, drizzles them with the maple syrup we boiled together in the spring, and grills them for a snack. My Beloved is in the Winter Garden, digging up potatoes, exposing nests of 'Eric' reds and 'Yukon Golds,' tossing the tubers onto the hay, where they dry in the paling sun. My Elder Son gathers up the crackling pods of black beans, pinto beans, and kidney beans in old wood-slat bushel baskets while my Beloved's Daughters glean windfalls from the orchard for applesauce and the New Son-in-Law chops celery for the freezer and some to be baked. I sit at the table, ticking items off my list, thinking how perfectly this holiday is timed to bring the family home, all of them, just when we need their helping hands.

We are lucky this year. There's been no frost yet, although we are already a week past our average first-frost date of October 5. Every day is a gift from the goddess of winter. The nights are cool and growth has stalled, but a few tomatoes still cling to the vines, and the wild grapes are luscious, a light bloom on the deep purple skins that stain our fingers as we pick out the stems before simmering the fruit for juice.

After solitary months of sowing, weeding, and hoeing, we're thrown into a turmoil of family harvest. The outside doors to the cellar are flung open as baskets of vegetables are carried down to their winter resting place. We could dig the dahlias and canna lilies, too, the red oxalises and gladiola corms, but I can't bear to part with them so long as there are flowers on the stem. But we pot up the rosemary and bay laurel, snipping the shrubs close, filling bags with fresh herbs for each of the children to take home.

"And tea. What about the tea?" the Elder Daughter asks, and so we gather the second cutting of lemon balm and spearmint while the Grand Girls pick the last few buds of chamomile and stems of catnip for their pets.

I wear my ladybug apron as I move from garden to kitchen, overseeing the picking and the storing, while we each take our turn preparing something for the meal. During a lull, the Grand Girls and I take the flower basket to the rose hedge, where we pick from the last flush—stems of white, pink, and fuchsia, and big, fat red hips that we arrange in bowls, buds, blooms, and berries marching in succession down the centre of the table.

Bending over the yellowing hedge with the two little girls, I feel a longing already for the innocence of planting, the possibilities of a seed.

But this is no time for reflection.

"Too bad we don't have a turkey," my Beloved says, eyeing the three chickens I am trussing for the roasting pan. "Those wild turkeys are so tame I bet they'd follow a trail of corn right into the oven."

He's kidding, but I would love to cook one of those birds for Thanksgiving dinner. Our first winter at The Leaf, a clutch of rough grouse were regulars at our bird feeder, waddling out from the woods in the first light of morning and again as the shadows lengthened. One day we heard a heavy thud against the window glass and found a lovely female lying in the snow, its neck at an impossible angle. I whisked it inside, cut off the head, wings, and feet, slid my hands beneath the skin and, with a sharp tug, slipped off the feathers like a coat. Gutted and roasted, it wasn't much bigger than a quail, scarcely enough for two, though the breast was plump, corn-fed from our own hand.

The turkeys would be tender, too. Our feeder is on their rounds of good dining spots. Some nights they roost in the pine trees above the kitchen and in the mornings we can hear them flapping to the ground, landing like great albatrosses on the roof. I've often watched them through the window, trying to guess which come from this year's crop of hatchlings and which are the tough old birds.

The eastern wild turkeys, the ones that strut across our yard, frightening the wits out of the chickens, are *Meleagris gallopavo silvestris,* the same species the Puritans encountered when they landed at Jamestown, and the same species most families are eating for dinner on this holiday. When the United States chose the eagle—a scavenger—for its national emblem, Benjamin Franklin proposed instead the wild turkey, "in Comparison a much more respectable Bird, and withal a true original Native of America. . . . He is besides, though a little vain & silly, a Bird of Courage, and would not hesitate to attack a Grenadier of the British Guards who should presume to invade his Farm Yard with a red Coat on."

There is only one red coat in our farmyard, and that is the old plaid my Beloved wears as he lifts the Grand Girls up to complete the first ritual filling of the bird feeders where the turkeys, grouse, and songbirds will graze over the next six months.

I pause to watch them, thinking, *Pay attention.* This day is a marker. One half of the year is over. The next is about to begin.

Someday I'll tell the Grand Girls about Beira, Queen of Winter, who made the mountains of Scotland where my mother's family came from. That country, like mine, has only two seasons, ruled over by a pair of women, Beira and Bride. The lovely goddess Bride presides over summer, which lasts from the first of May, called Bealltainn, to the first of November, Samhainn. During Bride's reign, Beira is beautiful, too, for as soon as the days begin to lengthen, the Queen of Winter drinks from the Well of Youth and is reborn as a lovely girl with hair the colour of yellow broom, cheeks as red as rowan berries, eyes as blue as a summer sea. But the transformation doesn't last. She ages quickly, and by now, in October, her teeth are red as rust, her hair white as snow, her cloak as grey as the clouds that are barrelling in from the north.

She was called the Old Woman, the Old Hag of Winter, and someone had to take her in. To decide who that would be, the first farmer to finish the grain harvest would make a corn dolly from the last sheaf. He'd toss the dolly into the field of a neighbour who hadn't yet brought in his grain, and so on, until the farmer who was last to finish the harvest had to care for the corn dolly—and Beira—through the winter until spring.

The story seems too dark for the day, the creatures too strange. Still, I ache for such rituals. I make do with my mother's recipes, my grandmother's dishes, my Beloved's grandmother's silver serving spoons, the red leaves ironed bright between sheets of waxed paper and strewn across the table for decoration, something I did as a child, and my children did, and now the Grand Girls do, too.

"Let's shuck some corn," I say to the Grand Girls, to lift me out of my reverie.

"Can we make corn dolls from the husks?" they say in unison, and I gather them to me in a tight embrace.

Dinner is almost ready. We're driven half-mad by the smells from the kitchen; we stand in line at the bathroom, eager to wash off the day's labours and begin the feast. What a lot we've accomplished! The garlic is braided and hanging in the cellar, the onions, too. The squash is mounded in wooden boxes in the cold cellar beside the bushels of potatoes. The drying beans wait in a corner of the Garden Room for an evening with nothing else to do. The beets and celery, the cabbage and zucchini are in the freezer; the herbs in their bottles; the tea in its caddy; the fruits, juices, and pickles in their gleaming jars in the pantry; the carrots and leeks in the garden under their mounds of hay. The Brussels sprouts, kale, late greens, and the last lingering tomatoes will see us through the final weeks, but this is it. The harvest is in.

We take our places at the long, cobbled-together table in the Garden Room, which my Beloved and the boy-men wrapped in plastic against the coming cold. We drop the woven blinds over the plastic and light the candles, then one by one, we bring in the dishes we've made, announcing each as we set it on the table.

"Everything comes from this place!" I exclaim in wonder.

"The salt?" says one Grand Girl.

"The pepper?" says the other.

"Not the milk," says the Younger Daughter.

"Or the wine," says my Beloved.

"Well, almost everything," I say, looking around at all we are about to consume that has come to us in packages, bags, and bottles. Flour. Sugar. Butter. Coffee. Then I perk up. "Maybe next year we could make dandelion wine and chicory coffee!"

I feel a sweet sadness wash over me as the conversation moves around the table, each of us declaring the thing that he or she is thankful for, which in the end is all the same. The food, yes, pulled from soil we've worked and known, and the company we keep, the love that criss-crosses the table like a cat's cradle. That we can still do this—still nurture ourselves and each other with what comes from this earth where we live—seems a most remarkable gift, and for that, I give thanks to the gardens that will wait for us in silence through the cooling night of winter ahead.

OLD KING COLE

My Beloved thrives on the shoulder seasons, the warming days of spring and the cooling days at the end of the year. I prefer the summer, with its massaging heat, but through the dog days, he's irritable, sweetening like the brassicas when there is frost in the air.

He pulls on his boots and heavy khaki jacket with pleasure. I reluctantly shed my flowered gardening shirt and get out my insulated gardening gloves, as grey as the lowering sky.

There's not much left in the garden, though some of the cabbages linger on. Cabbage, or cole—*Brassica oleracea*—is a large and diverse species that includes leafy kale, cauliflower, broccoli, kohlrabi, and Brussels sprouts as well as the heading cabbages you'd expect. The cauliflower and broccoli are long since finished, of course, and the kohlrabi, too. I used to pull up the heading cabbages before the first really hard frost and hang them, root and all, in the cellar; in midwinter, I'd peel the dried, yellowed, and faintly rotted leaves off the outside to reveal crisp green heads. I used to make sauerkraut, too, salting the shredded leaves in a tall, pale crock and setting a plate weighted with a stone on top until the juices bubbled and the cellar stank. All winter we'd eat cabbage soup and *choucroute* simmered with bacon and onions. But I had more family then, and more energy. Now my Beloved and I eat coleslaw through September and October, when it is in season. We buy our sauerkraut in glass jars from the market.

"It's too cold to garden," I complain as we trudge out to the Winter Garden, named because this is where we grow the food we put up for the winter—tomatoes, onions, garlic, potatoes, beans, corn, beets, carrots, broccoli, cauliflower, leeks—not because anything grows here through the bleak, white season. Even now, the place looks ravaged, the rows stripped of all but a few straggling weeds and late-season vegetables: across the front, hillocks of hay like grave mounds where my Beloved will unearth carrots and leeks after the snow comes; on the far edge, the orange beacons of two pumpkins the Grand Girls asked me to save for Halloween; and at the back, those last-ditch *Brassica oleracea*, the Brussels sprouts. Against the purple, snow-tinged clouds, the gnarled grey-green stalks carry on, studded here and there with knobs and globs, a palmy flutter on top.

"They look like aliens," my Beloved says, "or leftovers from the Pleistocene."

Brassica, as a genus, contains more food plants than any other. All its plant parts have evolved to the edible: leaves (kale and cabbage), stems (kohlrabi), roots (turnips), flowers (cauliflower and broccoli), and seeds (mustard seed and various oils). Such diversity suggests that brassicas have been around for a very long time. *Brassica oleracea*, with is multiferous subspecies, may be ancient. In fact, brassicas are so old that scientists are busy mapping the genome, sharing their findings at brassica conventions, where I imagine them referring to each other affectionately as "cabbageheads" and wearing buttons, BRUSSELS SPROUTS FOREVER!

It is strange to think of this rough, peasant green preening in the spotlight, but it's not the first time. Roman historians such as Herodotus, Cato, and Pliny the Elder documented the correct way to eat cabbage (with vinegar) and included it in a list of foods enjoyed by early Egyptians.

But I wonder: what do they mean by *cabbage*? We rely on words, yet language is constantly in flux. When they wrote *caulis*, did they mean the loose leaves of kale, the tightly bound heads we sliver into coleslaw, or a column of diminutive sprouts? Or maybe *cabbage* meant something altogether different at the time, a vegetable we wouldn't recognize if we saw it on our plates or smelled it cooking. Vegetative plants are shape-shifters: they reproduce prolifically, changing as fast as our tastes. Even Thomas Jefferson, who kept fastidious plant lists, left a legacy of mystery plants, including "sprout kale," which may be a kind of kale or a kind of Brussels sprout. We'll never know.

Even our Brussels sprouts look nothing like the picture on the packet that contained the seeds I planted. The illustration shows broad, vigorous shafts the size of cricket bats, uniformly thick top to bottom with perfectly rounded rosettes. Our stalks are hideously misshapen. Blown cabbages the size of golf balls stud the lower courses, decreasing to pinheads, then nothing but nubs of failed ambition at the top.

I follow the rules. I pluck the lower leaves through the summer, gradually baring the stems until there's nothing but a Samurai knot on top. When buds emerge at the leaf axils, I become vigilant. I pick the largest as they mature. There is nothing I can do about the hot spells that loosen the rosettes, make them flaccid and bitter, though I do give the plants good soil, digging in manure and dusting it with ash. But maybe I've been going at it the wrong way. Maybe it's water they're wanting. I think perhaps they'd like my garden better if it were a marsh.

I buy varieties called 'Prince Marvel,' 'Bubbles,' and 'Jade,' bolstered by the names. It was the French who brought Brussels sprouts to this continent, to Louisiana first. Thomas Jefferson, who is sometimes given credit for introducing them to North America, planted them later, though he, too, got his seed from Paris. Most of the world's sprouts are still grown in Europe—it's where they first appeared, nine hundred years ago, in the Belgian city of Brussels—though southern Ontario produces a thousand tons a year. So why can't I grow more than a couple of cupfuls?

"Why torture yourself? Why keep planting them?" my Beloved asks, a question I put to myself each fall during what passes for the sprout harvest.

Is it the seed packets that spur me on, those alluring photographs of what a champion sprout can be? Or is it my own stubborn pride? As October winds down, these stalks stand like wagging fingers in my looted garden, reminding me I've still a lot to learn.

"Because they keep me humble?" I shrug.

But there's more to it than that, I'm sure of it. I return to the picking and, bent low, look down along the row of gnarled plants. From this perspective, they seem monumental, as big as my Beloved, who crouches at the far end of the row. Suddenly, I think of the older women who have been my friends, women I didn't meet until their bodies were as misshapen and gnarled as these. Since I was twelve, I've always had a special friend who was seventy, fifty, thirty years older than I—women who pointed the way ahead.

"I think I grow sprouts mostly because they're old," I say at last, gathering up the basket, holding it in my arms. I look with new affection at these old crones. They live two seasons in my garden, sometimes three if the autumn is mild and the deer leave them alone, but no longer. We have a glancing acquaint-ance, these sprouts and I. Species passing in the night.

"I am a mere nubbin in *Brassica* history," I say to my Beloved as we head back to the house. He looks at me sideways, and I laugh. We've been alone in the garden all afternoon, yet I have that feeling I get when we travel or when I spend an hour in the company of an older woman friend. Buoyed, with a fresh perspective.

We're almost at the house when the first snow starts to fall. I don't mind it at all.

WHAT SURVIVES

"WE DON'T BELONG HERE," I say one morning as I look out the window at the puddles crackling with cold. My Beloved looks at me aghast. Such openings can lead to real estate listings and travel brochures. I am quick to reassure. "As creatures, I mean. We need fur."

"Humans are a savannah species." My Beloved especially likes to make this point at breakfast as we eat our mango with ancient-grain toast. "Look at the grocery stores, full of subtropical produce. And we keep our houses at an even twenty degrees Celsius. Obviously, we're trying to reproduce our ancestral home."

Maybe that's why I plant Kentucky coffeetrees and magnolias. In November, such attempts seem pathetic. Face it, I tell myself. Winter is coming. It is going to get cold, and then colder.

The garden buries its head in the earth. The delphinium tops die off, the hostas, too, until the garden flattens to an open plain. The plants that can't disappear altogether become skeletons, shedding their leaves to present naked limbs to the icy storms. In this, nature seems a perverse and unreliable guide. I bundle up.

I bundle up my precious immigrant transplants, too. Daisies and brown-eyed Susans may know how to cope with what's ahead, but the magnolias need my protection, the Emerald cedars and English roses, too. The ones that can't stand the cold at all—the rosemary, bay laurel, oleander, and mandevilla vine—are already inside.

"Why don't you grow what is adapted to the climate?" my Beloved says as he works the sumac poles out of the summer garden, where they served as ladders for 'Kentucky Wonder' beans. In winter they become the corner posts of my plant-huts.

"But the climate is changing," I argue. "I'm just thinking ahead."

I drive the stakes into the hardening ground around the rhododendron, the flowering cherry, the delicate magnolia, then wrap them with gunny sacks from a local coffee-roasting company. The Frisian heaps the cavities with leaves he brings from the village in orange plastic bags until, at the end of the day, the lawn is littered with what look like deflated pumpkins.

The plant-huts take us all afternoon. Through the next week, on sunny days, I wrap the cedars tightly, winding stout cord up the trunk to the peak, then down again, binding the branches until they look like skinny lances instead of elegant, elongated cones. We've been doing this ever since one spring, four years ago, when a burden of fresh, wet snow broke off a dozen cedar limbs and snapped our plump and stately Mountbatten juniper at the waist. Every year I intend to make plywood teepees for the ball cedars on either side of the front door, where avalanches regularly let loose from the roof, but in the end, I inevitably tie them up, too, and hope for the best.

The deep chill of dead winter isn't the problem. It's the kindness of a thaw that kills. Gardeners dream of winters where the snow comes early and deep, piling up steadily as the mercury slides, melting with calm resolve as the sun warms.

What we dread is the midwinter spike that liquefies snow over-night, warming damp-darkened bark until the sap rises. Then just when the plants are thinking spring, the temperature plummets, freezing everything in a flash. Tree limbs split; the tender shoots sent up from duped bulbs shrivel and blacken; buds tempted open by the heat are killed before they bloom.

And so I heap leaves over all the beds until they look like softly rounded graves. I lean boards against the south sides of young trees. I open bales of hay and set swatches against the leeks, building a cozy, knee-high archway that stretches down the middle of the winter vegetable garden like a miniature *allée*.

The Rosarian dropped by one afternoon as I was contemplating the roses, distraught at not being able to protect them, too. I'd read about kipping the shrub on its side and burying the stems to avoid winter dieback. I had already dug the trenches. But the canes were intent on resurrection. No sooner had I tamped the earth with the shovel and stood back to admire my work than the soil would crack and crumble and up they'd spring, refusing my clumsy burial.

"Lay an old rug over them," he said sagely.

And so now I keep in the shed a colourful array of carpet strips that I haul out every fall to stretch over the buried stems, to keep them to their place.

> My garden is transformed, a fine lady who has pulled on a rough old coat.

Within a week, my garden is transformed, a fine lady who has pulled on a rough old coat, Pygmalion in reverse. Gone are the flowing draperies of green and gold and scarlet; in their place, gunny-sacking, carpet bags, and coarse twine.

I rather like it. Enough of fashion: even the garden is in survival mode. As I clomp along the paths in my Sorels and old wool coat, I think, We make a pretty good pair.

LAST SONG

NOVEMBER ALWAYS COMES AS A SURPRISE—like the song that starts up after you think the CD has ended, a tune not on the playlist, an encore after the concert. A freebie. A gift.

In my mind, November is a cold, wet plod into winter. A time of sorrow for lost summer and regret for all that I have failed to do. The hostas I intended to separate and move. The trees I'd hoped to plant. The bulbs I bought but had no time to bury.

But then the penultimate month of the year arrives and the sun is warm! The soil is soft! The leaves are dry and crisp! How could I have forgotten November would be like this?

> How could I have forgotten November would be like this?

The thermometer registered double digits this week. I broke a sweat as I moved through my work. Is this heat normal? I check my Garden Book, where I diligently record the daily highs and lows, in temperature as well as in sowings and gleanings. The highs for the days of this week over the past ten years have ranged from plus fifteen to minus two degrees Celsius. No two years are alike. During the second week of November, sometimes it snowed, sometimes it rained, often it was sunny, and cloudy, too, but there is nothing to make me expect a November unremittingly gloomy and wet, too cold for a sensible person to be anywhere but settled by the fire with a book.

Why, then, do I remember the weather as uniformly bleak? Is my memory so skewed? Probably. A recent study in New South Wales discovered that on cold, windy, rainy days we remember three times more than we do when it is sunny. The study was carried out with shoppers who were asked to remember certain items in their carts, which may have no bearing on whether I remember to move the peonies, but even so, I find it curious. Rainy-day shoppers were less likely to have false memories, too: they rarely remembered objects that weren't actually there, unlike the sunny-day shoppers, who invented freely.

By this logic, I'm more likely to be right when I say, November is always rainy, than if I say, The anemones are gorgeous in the sun. According to the scientists, it has to do with mood. Rainy days put a person in a sour mood, and bad mood equals good memory. The theory is that feeling blue makes a person more likely to focus his attention on his surroundings as he glowers gloomily. (I use the masculine pronoun advisedly.)

I find it disturbing to think that I will remember my gloomy days more accurately than my sunny ones. This must be part of the same unhappy brain-quirk that makes us remember slights with such clarity, while forgetting the many small kindnesses we've been done.

Apparently, the effect works in reverse, too. When we're happy, we're likely to remember the weather as balmy. When we're sad, it's always raining. To skew matters further, a profound weather event can lodge in the memory so that forever after it is remembered as the usual, expected way of things. Perhaps if I searched through the meteorological records, I would discover a run of dark, gloomy November days when

I was five, the tail end of some hurricane or the low-pressure onset of some early winter storm. What happened, I wonder, to lay the pattern for Emily Dickinson's gloom?

And what about those who remember everything, every torrent, every sprinkle, every cloudless night and overcast afternoon? Such people exist. The first presented herself to memory researchers in 2006, and since then some fifty other memory mavens have come forward. They are blessed—or afflicted—with hyperthymesia, an acute autobiographical memory. *Thymesia* means memory in Greek, which strikes me as interesting, since the Romans recommended thyme to dispel melancholy, and with the melancholy the memory would go, too, no?

But here's the point: my gloomy expectations are misplaced. Even Dickinson would not call this November day the Norway of the year. The air is balmy, not a hint of the deep freeze that is to come. Time enough yet to get those tulips in the ground, to lift those sedums, bend the last rose canes to the earth and bed them down. Look: the Johnny-jump-ups are still blooming under the apple tree.

Listen. Can you hear it? The season's sweet last song.

AFTER THE FALL

LEAVINGS

THE SNOW IS A GODSEND. If it were blue or orange or even a uniform lime green, this absolute cover would be an irritation, an eyesore, but who can complain about a whiteness laid over the land like a feather boa?

"It's over," I say to my Beloved as I pull on my coat and boots. There is no sadness in my voice, only relief, and some joy at seeing again the bare bones of my garden, looking spare and fit.

Garden is no longer the right word. The place where I walk has reverted to itself, immune to my hovering hand, my fervent imaginings, the soil safe at last from my interferences. *But wait,* I want to say. *I'm not finished with you yet.*

Leftovers, that's what I'm looking for. Like a food addict standing at the door of an almost-empty fridge, I gaze over the stems and heads and boughs sticking up through the snow. I pull an old envelope from my pocket and pause over a juniper bush to pluck the berries. They're shrivelled and dusty with bloom, like old blueberries. My Beloved once saw juniper berries for sale in a gourmet shop for twelve dollars a pound. I push away the thought: I hate this tendency in myself to justify my pleasures. I like the berries; it's that simple. The look of them in the glass jar on the counter by the stove; the sound of them plopping into the stew; the taste of them, sharp and autumnal, under the smooth, meaty flavours of the sauce.

I check the sunflowers, but the blue jays have picked the heads clean. A few arugula stems are clacking in the wind; I open the pods and strip the seeds into another envelope. I find some errant poppies and scatter the seed like pepper over the snow. In the Kitchen Garden, I pick sage leaves, still green, thinking maybe I'll dip them in beaten egg white and sugar to make after-dinner cordials as holiday gifts for friends.

I am scavenging. Picking at the remains. This puts me, as my Beloved points out, in the company of vultures and hyenas. Once, though, scavenging was an honourable occupation, and may be yet again. A few years ago in Taos, my Beloved and I were offered lunch by earnest young men and women who had set up tables in the town square, serving bowls of soup made from vegetables they'd gleaned from grocery-store dumpsters. They called themselves salvagers. I like the sound of that. I'm a salvager, a gleaner in the garden.

And I'm not the only one. The chickadees and finches are here, too, pecking at the sedums and the brown-eyed Susans, the echinacea. Paths of voles and mice zipper across the snow.

I return to the house for my picking basket and secateurs. I have in mind a winter bouquet of the dogwood that has turned scarlet overnight, some milkweed pods, and a stem or two of bittersweet. The ornamental grasses I'll leave alone. I like the way they shake off the snow and brush against the faded sky as if wearing winter away.

It gives me such pleasure, this last round of plucking and picking. I don't need the seeds, and I can buy winter bouquets at the market. My friends are right: I am by nature a practical woman, but this salvaging has little to do with need.

I am saying goodbye, that's what I'm doing. I gaze over the gardens, only the vague shapes of the beds still visible under this clean white sheet, and as I do every year, I take my leave.

EVERGREEN

Up the path and through the Woodland Garden, on the way to my Beloved's writing cabin in the woods, is our Christmas tree cemetery: a single row of skeletons, unburied, limbs poking through the snow. It is obvious by the degree of desiccation, if not by their place in line, which were the first to die.

"We should have dragged them deeper into the woods," I say to him, sipping tea by his roaring stove. He works in his shirtsleeves; I wrap five layers against the drafts in the big stone house. "Or chopped them up for kindling."

"But then we'd forget," he says. In his mind, Christmas trees fall into the same category as chickens: if you're going to consume them, you'd better be able to kill them.

I sigh. It's time to murder another pine in honour of the season, and once again, the ghosts of Christmas trees past are haunting me.

This is a confession.

When my boys were young and we lived in the north, we'd trudge into the backwoods and chop down a tree for the holidays. Somehow, taking an axe to our own made it seem less of a crime. The forest was dense, the trees tall and gaunt. Often we felled a thirty-foot pine for the sake of a prettily shaped crown.

I consoled myself that the wind could have toppled that tree, or disease could have eaten it to a slow and disfiguring death. The trunk we left behind would become a nurse log for new saplings, I told the boys, though I turned aside at the sound of the chainsaw, at its rending and riving, the shrieks as the tree fell.

When we moved to the city, we turned to buying our Christmas trees. We'd drag them, bound in burlap straitjackets, out of their snow-fence compounds, strap them to the roof of the car, and drive them home through icy winds. Inside the house, we'd wire them to stand in the positions we wanted and load their branches with decorations.

How I loved those Scots pines! They brought the scent of the woods into the house, so that for a week, maybe two, when I closed my eyes and took a breath, the line between here and there blurred. I'd wake up in the morning to the fragrance of conifer and forget for a moment where I was.

Each year a different tree, though somehow they were always the same. Just as ladies in their manor houses once called all their maidservants Mary, we called all these trees, whether tall or short, fatly rounded or staunchly thin, the Christmas Tree. Roped with paper chains and popcorn, weighted with ornaments accumulated since childhood—miniature wooden sleighs, net angels, blown-glass birds, hammered tin stars and icicles, tiny sconces for the candles that we light now and then, a fragrant blaze—the Christmas Tree stood in the corner of one living room or another, freshening the air, feeding my memories, slowly dropping its needles onto the carpet.

By the time we moved back to the country, the children had moved on.

"We don't need a tree," we said that first Christmas, as though we had been putting it up only for them.

We lasted until Christmas Eve. "Just once more," we agreed, and rushed to the village grocery. The snow-fence compound was empty.

"We'll make a tree!" we exclaimed, undaunted.

These southern woods aren't home to many pines, but there are juniper shrubs in the meadow and cedars in the yard. I trimmed branches while my Beloved found a two-by-two that he drilled in a spiral. We rammed in the branches, trimming them to a shaggy cone.

"Not bad," we said, toasting our ingenuity with hot rum.

We would have slapped each other on the back, but it took both of us to hold up the Christmas Cone.

"Get the stand," I suggested, but our lumber trunk was too small.

"Maybe a planter?" my Beloved said, but the faux tree was too heavy. It toppled, bashing a dent in the branches, which made it seem more authentic.

He got out his drill.

"Not the ceiling," I said in a hushed voice.

"Can you think of a better way?" And so we hung the tree from a hook.

The next year we bought a tree. Two trees, in fact. The first, which my Beloved bought from the village compound, was a fir in its final throes. A trail of rusty needles led from the car to the library, where I'd cleared a corner.

"It won't do, will it?" my Beloved asked, seeing my tears.

And so he drove into town and bought a Scots pine that had grown for a decade in a field—ten years of sun and rain, hectic growth spurts followed by quiet periods of dormancy. Four thousand days of life so that my Beloved and I could gaze for a week on its swiftly fading loveliness.

"They're farmed," my Beloved said gamely, as we wrestled with our ghosts. "Like tomatoes."

He didn't buy his own argument any more than I did.

Last year, we purchased a pine seedling at a Christmas craft sale. High school students had raised the little trees from seed and potted them in bright containers to raise money for field trips. We draped a sliver of tinsel over its branches and kept it on a cool windowsill. Come spring, we planted it in the nursery bed along one side of the Winter Garden, where we coddle our foundling trees.

"A Christmas tree for the Grand Girls when we're dead and gone," my Beloved said.

I looked at him in horror. Would they really cut this tree down?

"Just kidding," he said.

It is only five days until Christmas now, and we've yet to decide on a tree. We missed the annual craft sale, but down our road a bit, a small spruce grows at the edge of the forest. Every December, someone decorates it with multicoloured Christmas balls. Recently, we passed a roadside tree hung with CDs that glinted in the whoosh of speeding cars. Nearby, another was strung with small coloured plant containers, dangling upside down like square bells: red and yellow, that particular burnt sienna, deep evergreen.

That's what I'd like in the house: a living tree, a full-size one that I could decorate, then take outside and plant. Instead of a graveyard, we'd have a Christmas *allée*. But the heat of the house would break the tree's dormancy. Even with care, moving it indoors for only a few hours at a time, it would almost surely die.

"Let's decorate the spruce in the yard," I say finally. "We could see it from the kitchen."

I imagine us stringing popcorn as we always do for our trees, but instead of tinsel and ornaments, we'll hang suet balls and tie the tips of branches with raffia bows.

"How much do you want to bet we find raffia in the birds' nests in the spring?" my Beloved says, breaking into a chorus of "Deck the Halls." I sing along, trying to think what else we could add to the tree, maybe cranberry-spiked mandarins and clusters of chunky nuts. There will be living ornaments, too: blue jays, cardinals, evening grosbeaks.

"No more lugging trees into the house," he says, warming to the idea. "No more needles in the carpet."

"No more resin on our hands," I say, thinking of Lady Macbeth.

"I could chop up those old trees for kindling," he offers on our way back to the house for lunch.

No, I think. Let's leave our ghosts lie a little longer.

THREE FRENCH HENS

I THINK OF A YEAR AS A GYRE, a length of something firm but malleable—a twist of clay perhaps—that moves through roughly the same course every twelve months. Where one length ends and the next begins, a flattened spot marks the join.

We're in that flattened spot now, that time between Christmas and New Year's that seems to belong neither to the year about to pass nor to the one on the doorstep. Daft Days, the Scots call them. This odd time-out-of-time coincides with the twelve days of Christmas, which start on December 25 and stretch to Epiphany on January 5, also known as Twelfth Night, the end of the reign of the Lord of Misrule.

By that counting, today is the third day of Christmas. According to the song, my Beloved should be giving me three French hens. What exactly is a French hen? I wonder. "The Twelve Days of Christmas" was first published in England in 1780, but it came from France, so a French hen may be nothing more than a loose translation, one that fits the rhyme and rhythm of the song.

But suppose they were particularly French breeds of hen. The three most common chickens in France at the time the carol was written were the Houdan, the Crevecoeur, and the La

Fleche. If my Beloved did indeed want to give me chickens for Christmas, he could be true to the song, for the same breeds are still pecking the ground today.

The Houdan, which is named for the city near Paris where it was bred, sports a crest as wide and downy as a parka hood. It also has a beard and muff and white-spotted black feathers, all of which combine to make it look a little doddering and dishevelled. The Crevecoeur looks like a refined Houdan, tall and prancing and jade black. Instead of that furry crest, beard, and muff, it has a spectacular spray of feathers rising like an iridescent black fountain from between its eyes. I wonder if Michael de Crevecoeur raised Crevecoeurs. He was the author of the first American best-seller in Europe, *Letters from an American Farmer,* which was published the same year as "The Twelve Days of Christmas." Jonathan Raban's book about America, *Hunting Mr. Heartbreak,* takes its title from de Crevecoeur's *Letters.* My Beloved would like the idea of a heartbreak hen. I'm not sure what he'd think of a La Fleche, which has an oddly forked comb that rises on its forehead like a pair of red devil's horns.

One year at Christmas, the Rosarian gave me not three French hens but dozens of chickens, or rather pictures of chickens pasted into a copybook, something he'd picked up at a local auction. The Houdan is there: it was exported to North America at the time of the American Civil War and appeared in the first edition of the *American Standard of Perfection,* the book that classifies and describes the appearance, colouring, and temperament of all recognized breeds of poultry, including chickens. The first edition, in 1874, included forty-one breeds; today's version has nearly sixty, categorized by traits such as "Houdan shape" and "La Fleche colour." The copybook has a dozen or so portraits, clipped, I imagine, from the *Family Herald,* watercolours of pecking hens and strutting cockerels, clustering chicks in reds and browns and blacks, speckled and striped, with hoods and muffs, and yes, beards.

I think of my staid flocks of Leghorns and Rhode Island Reds crossed with Barred Rocks. They could use an infusion of French *je ne sais quoi*.

"You have some catching up to do," I say to my Beloved. "A partridge, two turtle doves, a hooded hen, a fleshy hen, a heart-break hen. And tomorrow, four calling birds."

"Not calling birds," he says. "Collie birds. From *collier*: black as coal. Blackbirds is what I could be giving you tomorrow. A couple of brace of crows. You could make a pie."

I don't tell him that there are those who insist these are allegories: the partridge is Jesus; the doves, the Old and New testaments; the collie birds, the four gospels; the hens, that old triumvirate, faith, hope, and love.

There's a strange mix of the pagan and sacred in our house in these out-of-time, end-of-year days. I listen to Christmas carols—the Christian hymns of my childhood, contemporary Motown covers, Celtic solstice tunes—as I tidy up from our celebrations. It's unlucky to leave the trappings of Christmas hanging past Twelfth Night. Besides, the haggis in my bloodline reminds me that whatever remains undone by Hogmanay will plague me through the coming year. And so when the gifts are put away and the house returned to itself, I comb through my Gardening Record of the last decade, transferring my notes to myself to the brand-new book my Beloved bought for me: *prune all apple trees* (March 15), *harvest herbs* (July 30), *prepare compost bins for next year* (November 10).

The year coils before me in its predicable rhythms, slow to start through the winter, the pace quickening with spring. April

28 is the day I order the chicks that will keep us in roasted fowl through the winter. I pen the date into the new book and add a note: *Find three French hens.*

COLD COMFORT

I RARELY VISIT THE GARDENS IN WINTER. I look at them framed within windows, like paintings on the wall.

In my northern garden, the snow rose so high that I looked out on an unbroken expanse of white. It was a kind of erasure. I didn't mind it at the time, but now I'm glad that the snows don't fall so thickly on The Leaf. There is white, yes, but it is a perforated blanket. Plants poke through. The shrubs, of course—balls of boxwood and spikes of Adam's needle. The russet heads of 'Autumn Joy.'

One winter, my Beloved gave me a book, *The Gardener's Kalendar: The Works of Each Month in the English Garden*. The introduction offers instruction on laying out the plots—the Kitchen Garden, the Pleasure-Garden, the Nursery, the Orchard—on double-digging the beds, espaliering the apricots, constructing ha-has to keep the sheep from straying into the lettuce.

After these general instructions comes a month-by-month listing of tasks, starting with January, the time when, according to the author, I should be preparing hot-beds and planting cucumbers, asparagus, radishes, melons, and "small salleting," or greens. Peas can go into the ground now, and spinach. I should be pruning the grapevines, planting crocuses, anemones, and tulips. "This is also the proper season for laying turf where wanted, for making or mending grass-walks."

Did I mention that this garden was in the south of England and the year was 1777?

I will do none of this, of course, mostly because I can't. The weather is my master, and I submit to it with a smile. I gaze out the window, watching the chickadees pulling at seed heads and then flying to the fence, tucking their finds into knotholes while a black squirrel looks on, taking lessons in squirrelling.

When the sun is out, shadows pencil the snow. I congratulate myself on the procrastinations that left the monkshood uncut, and the alliums, too. After a snowfall, the sedum heads sport little pillow-caps. The naked vines along the fence are lined in white, drifted here and there with snow, random tracings like a Jackson Pollock painting.

The garden is lean and lovely: the ginkgo reaching every which way like a dancing Shiva; the flowering cherry with its nubby buds that will never bloom, but are pleasing anyway; the fuzzy magnolia; the big, fat pods of the Ohio buckeye; the simple line of the hedges; the shrubs, spare and elegant as Audrey Hepburn.

The garden is lean and lovely: the ginkgo reaching every which way like a dancing Shiva.

It's a black-and-white picture, colourized here and there. Ruddy dogwood stems. Jade boxwood. Red berries that dangle from the viburnum and cluster on the holly stems. The yellow grasses. Only the sky is blue. And the jays.

Oh, there are things I could do. It is warm outside today, not quite freezing. I could prune the suckers on the apple trees. Or trim errant shoots from the shrubs. Take the currants down to size. The fence around the chicken yard could use some attention.

But it's even warmer inside by the fire, and the view, as it is, is fine. My Beloved, already impatient with the snow season, likes to quote Shelley: "If Winter comes, can Spring be far behind?"

Let the work wait, I say. I'm happy in my respite, looking out at my garden, framed and lovely, wanting nothing.

BALANCING SCALE

EVERY SUMMER IN THE YEARS I LIVED IN THE CITY, I'd walk past a small diner on my way to the bank. Someone in the diner was a gardener, for there were crown-of-thorns in the window and, on the sidewalk, a row of oleanders with hot pink flowers, their little cardboard price tags nodding come-hither in the sun.

"You want to buy?" the woman from the restaurant would say, and I'd shake my head sadly. I did want one of those oleander shrubs, wanted to breathe in that fragrance and be transported to my youth, to the soft, musky scent of the fragrant hedge I used to pass on my way home from school. But I resisted. *Nerium oleander* is one of the most toxic plants on the planet. The sap contains powerful glycosides that speed the heart to a breakneck pace. The bark is laced with rosagenin, which acts like strychnine, causing convulsions. A handful of leaves will produce severe gastrointestinal distress in an adult; a single leaf can kill a young child. Even when it's old and dried and dead, this plant is deadly. I read once that a group of Napoleon's soldiers died when they used oleander sticks to roast meat over a fire.

As long as there were children in the house, I forswore the lovely, lethal oleander, but one day, when our nest was empty, I succumbed. The shrub was about four feet tall, the leaves glossy, dark lances on stems that bent gracefully under the burden of the fuchsia flowers at their tips. How could something so beautiful be a threat?

All summer, the oleander sat on our back deck. I urged it on, feeding it generously, looking forward to the day I would take cuttings and start an oleander hedge, one the Grand Girls might walk under, gathering their own memories.

The shrub grew quickly. I stripped the lower half of the three main stems and wove them together to form a trunk. The top feathered out like a small tree. In the fall, I repotted it and moved it into the house, where it continued to bloom, though in the absence of strong sun, the colour faded to a pale, prophetic rose.

Laurel rose. Rose-bay. French rose. Desert rose. St.-Joseph's-flower. I got to know it by all its names.

It was around Christmas that I noticed the leaves. They drooped with a kind of post-bloom depression. They'll perk up, I thought, and I fed the plant a little more. The leaves began to fall. I looked more closely, poked at them with my finger. The topsides were sticky and black; the undersides spotted with what looked like pregnant sesame seeds. Little growths clustered in the crevices of branches, at the junctions of leaf and stem. Fresh new stems were lumpy with tiny tumours.

I shuddered. This was no disease. It was soft brown scale. I had been introduced to this insect on a jade plant that I carried from university residence through a string of apartments to my first house. Over the years it had grown to the size of a peony shrub, countless small branches bifurcating every which way, hundreds of fat, fleshy leaves, and at every junction of leaf and stem, the sucking hordes.

"Discard heavily infested plants," my wise old garden encyclopedia advised. I threw the jade plant into the snow.

I am slower to take such drastic measures now.

There are more than seven thousand species of scale, belonging to three basic types: mealy bugs, which have no shell; armoured scale, which secretes a tough, waxy covering separate from the insect's body; and soft scale, which secretes a softer covering that remains attached. There are some eighty-five species of soft scale in the United States, probably fewer than that in Canada. It doesn't matter: you need only one on a plant to understand the kind of devastation it can cause.

I should have known. For weeks, I'd been ignoring an odd stickiness on the floor under the plant. Soft scale consumes so much plant sap that it excretes a stream of sweet liquid called honeydew, which coats the leaves and stems and drips onto the floor. Honeydew is the perfect medium for sooty mould, a dark fungus that spreads over the leaves, obstructing photosynthesis and choking the plant.

The lumps along the leaves and stems were female scale. Eggs are laid and hatched under the female, and when the nymphs are mature, they crawl away to a juicier spot and start sucking, losing their legs and antennae and growing a shell of their own. The males, which are microscopic, sprout wings and fly about, though they don't travel far either.

They don't have to. In the warm, dry world of a living room like ours, scale can go through up to six generations a year. The numbers don't bear thinking about. These insects live to sink their mouth parts into plant flesh and suck the life juice out of it. I understand that. But if oleander sap is lethal to humans, cats, and birds, why doesn't it kill something the size of a freckle?

Oleander sap might not have been up to the job of murdering scale, but I was. I washed the leaves with soapy water, prying off the beasts with my fingernail. Within a month, more appeared. I washed and washed. When the Rosarian invited me to a meeting of the local garden club, where the speaker was an expert on indoor plants, I went armed with my question:

"How do I rid my oleander of scale?"

The woman smiled a pitying smile. "Alcohol," she said, without conviction. "And elbow grease."

I bought a mighty jug of spirits and set to work, wiping down each leaf, both sides. I picked off each individual scale with a Q-tip. I draped a sheet over the shrub, isolating each part as I worked, to avoid contamination. The treatment took three days, but at the end of it, the oleander was squeaky clean, the leaves bright green. As a precaution, I scraped away the top few inches of soil and added fresh. I set the plant in a south window of the library, where in winter I like to stand and look out over the snow-covered garden. I took to running my fingers through the leaves in random spot checks.

Whole months passed without my seeing a telltale lump. I grew confident, then cocky—until, in March, I saw one. By April, when the plant was setting buds, there were dozens. I kept a spray bottle of alcohol and a store of Q-tips handy, but I couldn't keep up. I gave the whole shrub a full treatment before I moved it outside for the summer, then hoped that natural predators would take care of any I'd missed. In the fall, the oleander got another treatment before coming indoors for the winter. By January, it was infested again.

I scrubbed it down. I read everything I could find, followed every rule like a nun in training. Don't over-fertilize: scale loves a plant with high levels of nitrogen. Be sure to moisten the leaves:

scale thrives in an arid environment. Check the plant often: I combed through its leaves like a mother searching for nits.

All I ever managed was control, never a cure. That the oleander survived at all seemed a triumph. Through the summer, it blazed with blooms. It grew as if there were nothing holding it back, until its canopy was taller now than I am, its woven trunk as thick as my wrist.

I kept my ears open for other methods, some key bit of information that would help me deal a final killing blow to the scale without poisoning the plant, or my Beloved and me. Cigarette butts. Ground-up chrysanthemums. Insecticidal soaps. Tea tree oil. Last summer, environmentalists were touting Listerine mouthwash as an effective spray against mosquitoes, so I thought, *Why not scale?* I sprayed the plant heavily, waited half an hour, then hosed it off. The scale turned dark. The insects were dead. Hallelujah! Could it really be so simple?

Apparently not. Here it is the end of January, and right on schedule, the scale is back.

What more can I do?

One source tells me to attach two-sided tape to the branches or isolate an infested leaf in a baggie so I can tell when the crawlers are on the move, then do a treatment quickly before the adults are plump under their shells. Another suggests I invest in a counter-infestation of *Metaphycus helvolus*, a scale-loving wasp. Or ladybird beetles. Already our house is alive with beetles every spring, and they don't seem to relish the scale at all. Wrong kind of beetle, I suppose.

One blogger suggests using facial sponges instead of Q-tips. Maybe I should change my brand of alcohol. Use rum. Or Scotch. Or maybe the pinga my friend brought me from Brazil.

Maybe I should stop buying plants off the sidewalk.

Maybe I should just toss the oleander into the snow.

I recoil at the thought. When we moved to The Leaf on a bright February day, leaning against the garbage by the road was a fully grown hibiscus tree that the previous owners had thrown away. I wept to see it, frozen solid, the leaves black with cold, though I'd done the same years before to my jade tree. But the oleander? Never! Is that because it is more beautiful to me? Or because I see myself in that shrub, flawed and demanding, but worth loving just the same?

I'm not about to give up. It isn't a battle anymore, just another midwinter ritual. My Beloved helps me drag the oleander into the shower, where we both strip down and go at the leaves with scrub brushes. I put the shower head on pulse and direct the heavy stream on the plant.

"Maybe that's what it's been missing," I say. "A torrential downpour, like we used to have every afternoon in Brazil."

When the plant is clean, I'll spray the leaves with insecticidal soap, let it rest, then wash it off one last time. Every week through the winter I'll do my leaf-checks. I'll think I've destroyed every last flake of scale, and back it will come again. I'll despair that the poor plant will die of all that infernal sucking, but it won't.

It hasn't yet.

MARY HART AND THE COUNTESS VANDAL

WINTER. THE DIGITAL OUTDOOR THERMOMETER has given up and simply registers OFF. In the lovely brass instrument by the Garden Room door, positioned precisely so that neither my Beloved nor I has to leave the house to read it, the mercury has shrunk to a mean, glinting ball. It is minus forty degrees, the point at which the world's competing systems of measurement call a truce. Fahrenheit, Celsius: what does it matter? By every standard, it's cold.

Cold and white.

Maybe this is why gardeners in the northern half of the continent go a little crazy when the seed catalogues arrive. These bright little bundles of promises are cunningly timed to appear in the mailbox just as I despair of ever seeing green again. Short-sighted human that I am, by midwinter I believe wholeheartedly in this polar landscape, the enduring inevitability of black, white, and grey. Red tomatoes from this frozen earth? Golden daffodils? Impossible.

I thumb the glossy pages like a penitent still in the desert, gorging myself on rich descriptions: *long glossy leaves with a tangy, juicy flavour . . . sweetly scented with dark chocolate cones . . . plum petals with creamy vanilla centres . . . crisp and white with a spicy aroma . . . cascading tresses of rich orange fruit.* I am a gardener, prone to visions, especially in January, when I grasp at the catalogues, plant-starved.

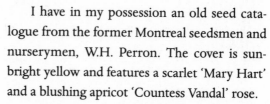

I am a gardener, prone to visions, especially in January, when I grasp at the catalogues, plant-starved.

I have in my possession an old seed catalogue from the former Montreal seedsmen and nurserymen, W.H. Perron. The cover is sun-bright yellow and features a scarlet 'Mary Hart' and a blushing apricot 'Countess Vandal' rose.

W.H. Perron—"The Reliable House"—was located in the heart of Montreal, just south of the corner of St. Laurent and Dorchester, though it grew its plants across the Rivière des Prairies in Laval. Perron was also a mail-order seed house. By 1930, it had catalogued its holdings and was offering them by post to gardeners who couldn't come in to shop at the store. This catalogue was its fifth, published in 1935. Packets of seed cost 15 cents. Hydrangea roots were $4 a dozen. Amaryllis bulbs, $1.50 each. The back pages offered the usual plant food, window boxes, 'Dogzoff,' and tree tangle-foot, as well as more arcane grave-flower holders and asparagus bunchers.

Compared to the productions of today, this catalogue, at first glance, looks a little dull. Inside are 140 pages of fruit, vegetables, and flowers, the listings thoroughly if not lavishly illustrated: a pyramid of perfect 'Snowball' cauliflowers, a truss of leeks, a vine dangling with fat peas, all shot in richly toned black-and-white. The edges of the pea-vine are blurred into the page so that the pods seem to dangle right at your fingertips. Most of the vegetables are shown as ink sketches or lithographs: 'Mammoth Mangels' and 'Irish Cobblers,' turnips and 'Golden Summer Crookneck' squash, all finely lined and cross-hatched.

The flowers, for the most part, are shot against black back-grounds, giving them a formal, painterly look. In fact, I'm pretty sure some of them *are* painted, or maybe the photographs were hand-coloured.

Only the covers and the sheets of heavy paper that divide the catalogue into thirds are printed in colour. These inserts feature giant dahlias, oversized Swiss pansies, large, fluffy ruffled St. Joseph petunias, and a field of multicoloured gladioli. Four enormous 'Dorsett' strawberries dominate the back cover, along with a tumble of 'Latham' raspberries of an intense, impossible red.

It is odd how this old-fashioned catalogue takes hold of me. When I return to this year's catalogues, they seem unattractive by comparison, even though they are generously illustrated with full-colour photographs on every page. Shouldn't photographs be an improvement over drawings and paintings? Not an artist's representation of a carrot or zinnia, but the very thing itself. The squashes are clearly squashes. The morning glories are blue and climbing. The peas are hanging on the vine, and sometimes splayed on the ground, too, away from the masking play of shadows on the leaves. Pictures taken in a real person's garden.

Still, after thumbing through Perron's catalogue, I feel dis-gruntled with the newer versions that have been arriving on my doorstep. Their glossy photographs are like television: they leave nothing to the imagination. The old lithographs are more like radio: they lack colour and context, gaps I have to fill in for myself. Unhinged from reality, these images might come from anywhere—and they fit anywhere, too. And so I imagine the plump pole of 'St. Fiacre' beans in a corner of my Winter Garden by the asparagus, the bouquet of 'White Paris Cos' lettuce in the Kitchen Garden by the sprouting onions.

"It might be like what Janet Malcolm says about writing," my Beloved suggests. "People believe fiction more than they do nonfiction, because with nonfiction, they can always say to themselves, Well, that's one side of the story."

Precisely. I look at those glossy photos and I know that *that* lettuce will never be my lettuce. But with a sketch, I think, If I dig in enough manure and hand-pick the slugs, maybe . . .

Sometimes, I admit, I bypass the catalogues altogether and buy my seeds at Canadian Tire, at their end-of-season, four-for-the-price-of-three sale. I like touching the packets, feeling their lumpy cargo, carrying them in my arms to the checkout. No lists. No alphabet to pore through. No high-resolution images to compare my efforts to. I twirl the display stand, stopping it here and there to take a closer look at whatever catches my eye. The shopping is free-form, a complete surprise, like navigating without a GPS.

The first catalogues arrived in December, and they continue to appear even now, in February. I got another one today. My Beloved collects them from the mailbox and stacks them by my chair beside the fire. When all the usuals are here, come home like family for a midwinter gathering, I lift my old Perron catalogue from the shelf and settle on a Sunday to order seeds for the coming garden year. I can't order the 'White Swan' peony or the 'Glorious Gleam' nasturtiums, of course, but I read the Perron first. It puts me in the mood.

This morning, as I cruised the Internet, looking for new nurseries, unusual catalogues to request, I discovered that W.H. Perron still exists, in a way. Wilfrid-Henri started the company in 1928. In 1992, W.H. Perron bought out Dominion Seeds, and two years later, Perron itself was bought out by White Rose Crafts and Nursery. That same year, Perron's company changed its name to Norseco—North Seeds Company—which is a plant-and-seed wholesaler. The venerable old name was preserved in its francophone consumer branch, HortiClub W.H. Perron; anglophones have to shop at the Dominion Seed House.

At the Dominion Seed House website, I see nothing at all about Wilfrid-Henri. Dominion also got its start in 1928, a good year, apparently, for seeds, though it seems peculiar to me. I wouldn't have thought there would be an upswing in gardening at the height of the twenties. Two years later, maybe, when the Depression had people digging in the dirt again. Perhaps both seed houses were prescient.

In any case, it was the Harding family who started the Dominion Seed House in Georgetown, Ontario, where its main office is still open from 8:00 a.m. until 5:30 p.m. at this time of year. Norseco isn't mentioned.

I scroll through the offerings. No 'Mary Hart' rose. No 'Countess Vandal.' No 'Snowball' cauliflower. No 'Dorsett' strawberry or 'Latham' raspberry, although it does advertise a raspberry called the 'Perron Red.'

I click "Add to My Shopping Basket." A new garden season has begun.

A DRY SEASON

IN SUMMER, MY EYE IS TUNED TO THE SKY, to the wisp of horsetail that foretells a shift in weather, to the billowy build of cumulus that spells rain. I am always on the lookout for those mild, cloudy days conducive to transplanting or the bright, clear stretches of sun the onions need to dry. I think of weather as a friend to my garden, though a quixotic one. I never know what it will do next, and so I keep a close watch.

In February, I hardly look up. I may be aware that it is snowing or sunny, but only vaguely, and in a wholly personal way. A lozenge of warm sunlight on my desk. An unsullied swath of snow beyond the window. Dreary runs of cloudy days. Especially this. In the north, the winters were brilliant with sunshine, but at The Leaf, the winters are long and shrouded. I feel deprived, disgruntled, my spirits as low as the clouds.

In winter, I become the kind of person I complain about the rest of the year. "Listen to those city media guys!" I rail to my Beloved every summer. "Complaining about one measly rainy day when the lettuce is so desperate for water. Don't they know where their food comes from?"

Now I moan, "Why can't the sun shine?"

But it is shining, at last. Shining on and on. This morning, the radio declares a winter drought.

"By this time last year, almost six feet of snow had fallen," the deep voice of the announcer says. "So far this winter, there has been less than half as much."

Outside, a skiff of snow skitters across almost-bare ground. I am jarred from my winter cocoon. What is a winter drought?

I remember drawing The Hydrologic Cycle in Grade Five science class—a puddle of a lake with red arrows waving up into a clear sunny sky, clouds forming to the right, letting go straight blue shafts of rain that ran in a stream down a green slope back into the lake. I loved the neatness of the system, the predictable, circumscribed loop of it. The point, of course, was that a finite supply of water is endlessly circulating the earth, sometimes as vapour, sometimes as liquid, and sometimes as solid snow and ice. *Finite*: that was what I was meant to learn. Instead, I left Grade Five feeling confident that an invisible pump was working away, reliably sucking vapour up into the sky and letting loose precipitation in equal measure—a tight watery loop that would spin forever in my backyard.

It never occurred to me that water might be sucked up one place as vapour and deposited somewhere else as rain. Or that the cycle would continue through the winter, when the lake is rigid with ice and no streams flow through the snow.

The climate where I live is known as "humid continental," which makes me think of April in Paris. The humid part comes mostly in spring and fall. Summer and winter are drier, with February the low point. To be fair, at that time of year, water arrives most often in the form of snow. Most years, snow covers the ground from December to March; a third of it comes in February. I love to see that thick insulating cover. When it melts in midwinter, as it almost always does, at least once, I despair for the plants. Bulbs nudge up through the soil, and the leaves of the primroses perk up as if spring has come. Their clocks are set not to minutes and hours but to light and warmth, and they are easily deceived.

This year, February is cold, but we've had less than 10 percent of the usual snowfall. Day after day, partly cloudy skies, nothing but a few stray flakes gyrating in the breeze.

"No weather today," the announcer declares.

No weather? I sit up in my bed. No weather of interest to drivers and skiers and kids praying for snow angels, perhaps, but I'm fully awake now and I'm talking to the radio. "What about the plants? The soil? The trees?"

The trees will be fine, I tell myself. The shrubs, too. Last summer was cool and wet, so the soil was deeply moist before freeze-up. Plants that lose their leaves in winter don't transpire much, and if they've been in the ground awhile, they have root systems extensive enough to replace whatever is lost.

But I worry about the shallow-rooted and the newly planted, especially that cedar hedge we put in late last spring. Evergreens lose moisture through their foliage all winter long, particularly on bright, sunny days, when the sun warms the narrow leaves, releasing vapour to the cold, dry air.

I worry and I worry.

The root balls of those cedars will still be small, and besides, the ground is frozen. The roots won't be able to take up moisture and I can't haul out the hoses or turn on the outside water taps without putting the plumbing at risk. I could carry buckets of water to the new cedars, and I would, if there were only one or two, but the hedge is more than a hundred feet long.

"It's only February," my Beloved reminds me.

He's right. We could yet see the snowfall of the century. March could come in like a rain-soaked lion, bringing water when the shrubs and trees need it most, just as they are breaking dormancy and starting to grow.

I wish I knew for sure: is this a winter drought or just another dry spell? If it were July, I'd thrust my shovel into the soil to see how far down the moisture lies. But it's February, and the soil, even though it's bare, is locked up tight, keeping its secrets to itself.

"Stop thinking about it," my Beloved says. "There's nothing you can do."

"I suppose," I say, gazing out the window at the withered leaves playing tag among the dry sticks of veronica and sedum. I have no power over this. The climate is changing: the summers are hotter, the winters are milder. Strange beetles are moving north; the ground won't freeze hard enough to kill them off. And now this drought.

I scan the sky for snow clouds, hoping for cover for the parching soil. If I could be out there, mulching and tending the plants, I would at least feel that I was doing something. As it is, all I can do is sit at my window and wait.

STIRRINGS

HOOKED

I CAN'T WAIT ANY LONGER. The ground is still winter-locked, but I'm ready to break out my garden tools. I pull open my seed box, an old oak case that was home to file cards in some long-ago office, stuffed now with seed envelopes arranged according to the date the seeds go into the ground: beans outdoors in early June; melons indoors in late April; spinach in the garden at the end of March, as soon as the ground thaws; tomatoes and peppers under lights in the house in mid-March, and in February, impatiens, which is how I feel right now—impatient for growing things.

Impatiens seeds are as fine as the coloured sprinkles I scatter on cupcakes for the Grand Girls. I've done this for years, pressing seeds into soil, but still I find it hard to believe that each tiny pellet will become a full-grown plant. The puddle of seed in the palm of my hand is enough to populate the broad bed at the base of the apple tree, with plenty left over to spot colour through the Woodland Garden, the Shrubbery, and around the Grand Girls' playhouse.

I plant too much; I always do. I collect too much, save too much, sow too much. My enthusiasms overtake me. I'd plant too early, too, if I could, but like a drunk who hides his bottles behind books on the library shelf, I stash the planting pots along a wall that is obliterated every fall by our winter wood supply, in a place carefully calculated not to be exposed before the middle of February. This year, my plan backfired: the weather has been mild, which means the woodpile has diminished slowly. My pots were trapped, my seeds unplanted, my fingers becoming twitchy until finally, on the last Sunday of the month, in a frenzy, I toss the firewood aside.

I plant too much; I always do. I collect too much, save too much, sow too much.

I haul the pots up from the basement, together with the bag of professional potting soil I bought last fall and the child's wet-play tray I use for mixing. If I were more of a purist, I would bag my own screened compost in September and sift it together with sand and perlite, but the chemistry confounds me. I've never been able to produce a mix sufficiently light and sterile, and the truth is, I don't care enough to learn. I don't make my own tools, either. Some things are best left to the professionals.

I lay out my supplies on the kitchen counter. Scoops of various sizes. A shot glass full of toothpicks. Plates of a certain colour: one red, one white. A grease pencil sharpened to a point.

I always plant on a Sunday. I water my indoor plants on Sundays, too. I grew up in a family that dressed for church on the seventh day, then gathered for the Sunday roast, followed by a quiet afternoon of reading and playing games. Scrabble. Snakes and Ladders. Parcheesi. Small rituals to mark the week, an end and a beginning, too.

"Playing in the dirt again?" my Beloved asks, coming into the kitchen as I pour a pitcher of warm water into the crater at the top of the mountain of soil in the tray. I sink my hands in, working in the water, kneading the earth. I close my eyes and smile. He kisses my cheek. "Enjoy."

Plants at our house have had all manner of beginnings. My first seedlings came to life in egg cartons on a windowsill. For a time, I used a brass press that formed the damp potting mix into cubes with a depression in the top where I'd press a single seed. The squares sat in a tray and soaked up water on all sides, so that the young roots spread evenly and robustly. But the trays disappeared in one move or another, and although it is one of those things I look for at every yard sale, I have never found a decent substitute.

Mostly, I use recycled nursery pots, thin black plastic divided into four or six or eight cells that I scrub before I plant. I pick them up at the dump, from neighbours and friends. I like to start small, transplanting once or twice as the seedling grows. I tell myself that planting seeds into a pot that will accommodate a full-size plant is like laying a baby in the middle of a king-size bed, but the truth is I do it, too, for the pleasure of holding the tiny stems, cradling them pot to pot.

I pour the seed onto a plate—white for black seed, red for white seed—then I wet a toothpick on my tongue and pick up the seeds one by one, tapping them into place, one to a pot. The experts advise planting three to a pot, but I can't bring myself to pinch off the weakest two, the way I'm supposed to. Instead, I lift them to other pots, ending up with three times more than I need. Better to limit my urges where I can.

Each seed contains a store of food and a dormant embryo, which, when exposed to the right balance of light and warmth and water, will sprout into a plant. In my boundless optimism, I expect every seed to sprout, but a plant is more pragmatic, producing thousands of seeds in the hopes that a few will survive to carry its genetic message into the future. Seed may be blown on the wind, fall on barren ground, slip into a dormancy that might never be broken.

Most of my seed is fresh, gathered from last year's garden, but the season was bad for impatiens. Whenever I went looking, the touch-me-not stems were either bare or the pods had already burst, scattering seed on the ground, where it would succumb to winter. It's a good thing I never use all the seed I collect and never throw out the leftovers, which means I always have seed, in some cases envelopes that date back as long as I have gardened. I imagine myself in some apocalyptic future, doling out seed to neighbours and passersby, restocking the blasted earth.

The delusions of gardeners.

But it could happen. Archaeologists excavating King Herod's palace near the Dead Sea recently found some seed from a date palm, *Phoenix dactylifera*. It sprouted, producing a palm carbon-dated at about 2,000 years. And in China, some ancient Asian water-lotus seed (*Nelumbo nucifera*) germinated after 1,200 years. I've heard that seed from an Arctic lupine (*Lupinus arcticus*), found in a lemming burrow in the tundra, germinated and flowered after 10,000 years of dormancy. Surely my three-year-old seed will sprout.

All afternoon I pat warm soil into pots and press a seed in each, labelling as I go. I used to take my cue from the nurseries and stick little tags in the pots, but they'd get knocked out of place or the writing would fade and I wouldn't know what I had until the plants bloomed. Now I write the name of the plant on the side of the pots as I work.

Wash. Pat. Press. Write.

"Why do you do it?" my Beloved asks as he sets up the tiers of grow-lights. I've been at it for hours. My nails are dark with potting mix and I have daubs of soil on my nose where I push up my glasses. He can tell my back is sore: I'm slumped awkwardly over my wet-play tray, squeezing one last batch of mix. I started with impatiens, and though I know it's too soon, I've moved on to tomatoes and peppers, celery, celeriac, leeks, some early lettuce. "You could buy all this for a few dollars."

He's right. I could march into any nursery and bring home trays of exactly the right number of seedlings already hardened off, standing tall, ready to go in the ground. No months of careful watering and shifting pots under the lights to counteract the lean of stems. No anxious vigil for damping-off.

"I could," I agree. "But I can't."

I hand him trays that he slides into tight rows under the lights, paying close attention to the names I've written in white grease pencil on the black sides of the pots. This is just the beginning. I'll be planting every Sunday for the next few weeks, working my way through the brassicas to the melons and fall flowers. By the time I am finished, every square inch of free space in our house will be under lights—the dining room table, too, covered with pots nursing seeds that stir to life in darkness.

"Why not?" he says, straightening after his labours. I lay a finger on the soil. It has cooled to the temperature of the lights. It's a waiting game now, as the embryo absorbs the food stored within the seed. In a month or a week or a few days, depending on the species, an embryonic root will swell, breaking through the seed coat to poke downward; a soft plumule will push up. In some plants, the first leaves, the cotyledon, will remain underground, but in others, the ones I wait for, the "hook" of the plumule will push up through the soil, lifting the cotyledon into the air. Sometimes, the seed coat is still attached to the greening leaves, and I grasp it gently, slipping it off so the leaves burst apart and spread with what looks like a sigh of joy.

That pleasure is still ahead of me. For now, I attach the timers one by one, setting them to the length of a summer's day. From this moment on, I'll awake to the glow of daylight in my kitchen. No matter that the sun isn't yet up. I'll peer into the pots, into the loamy dampness of this soil, and wait.

> My heart lifted then as it will again, at the sight of those fragile green hooks lifting the leaves into light.

When the eldest Grand Girl was born, I was there to see her first breath, hear her first cry, see her first look at the world. My heart lifted then as it will again, at the sight of those fragile green hooks lifting the leaves into light.

I turn to my Beloved. "And miss all this?"

LOST IN NEVER NEVER LAND

I NO LONGER FALL PREY TO THE Fifth Avenue rhetoric of garden catalogues that offer pumpkins "prized in France for their yellow-orange flesh"; grasses whose "silky blooms begin as a deep rose, unfold to delicate pink, and shimmer in the autumn sun." As winter drags through its last weeks, I often return to those glossy pages, aching for spring. The colourful pages fairly drip with succulent vegetables, fruits ready to burst from their branches, flowers to feast the eye. Cheddar cauliflower? Rainbow carrots? White eggplant? Blue potatoes? Black tomatoes? Yes, I moan in ecstasy. Yes. I want it all.

But I am disciplined in my ordering. Long ago, I developed a strategy, a system not unlike the one I used when I was young and poor and feeding four on an artist's paltry wage. I make a list. I stick to it. I allow myself one, maybe two, at the most three deviations.

Never shop hungry is the other rule, but that's impossible with a garden. By March, I am starved for growing things.

Where I fall off the wagon, predictably, spectacularly, where my good intentions are about as much use as a dibble in clay, is at the nurseries that open their doors sometime around the first day of spring. I am fairly safe at the place I've been going to for years, where I trust the owner when she says no way that rhododendron will survive a winter on my sweep of land. But I am lost the moment I step into one of those fenced-in garden kiosks that pop up like forgotten bulbs on the parking lots of hardware chains and grocery stores, nurseries where no one knows me well enough to stay my greedy hand. Entering those leafy domains, I am transformed. Suddenly, I believe what I read. *A black buddleia hardy to zone 4.* Yes! *A hydrangea that blooms endlessly through the summer.* Why not?

My gardens at The Leaf sit squarely in zone 5. That doesn't mean that I can buy any plant labelled *Hardy to zone 5.* Many of our plants come from the United States, where the idea of those striped zone maps originated. That country bases its zones on average annual minimum temperatures. Canada produces its own plant-hardiness map that takes into account more variables: the amount of summer rain, the number of frost-free days, wind speeds, snow cover—all those things that experienced gardeners know can mean as much to plant survival as a number on a thermometer. The difference can be significant. The Leaf, for instance, sits in zone 4a on the American scale. Plants grown south of the border and marked *Hardy to zone 5* will not likely survive the winter here.

They are the Paris Hiltons of the plant realm, coddled and spoiled.

I forget all this the minute I enter a nursery. The plants are so blooming gorgeous! Somewhere deep in my cranium I know they were brought to this lush state of perfection in cozy greenhouses that deliver precisely the right combination of water and nutrients and pest control at exactly the moment it is needed. They are the Paris Hiltons of the plant realm, coddled and spoiled. They know nothing of the real world of intermittent rainfall, earwigs and slugs, and soil the consistency of brick. In my garden, their leaves will wrinkle in disgust, their stems will grow gaunt, their flowers will droop in dismay, Cinderellas whisked away from the ball.

I know this, and yet I am seduced. I buy the Japanese maple because I love the cascading drape of those scarlet leaves. I buy the witch hazel because it promises yellow blooms in January, and what winter-weakened soul wouldn't fall for that? And the rhodo: I succumb to the rhododendron because I once toured a British Columbia garden where a couple had spent a lifetime travelling the world to bring home hundreds upon hundreds of shrubs, some knee-high, some bending over us like deacons, with fat scarlet blooms. There are hardly enough adjectives to match all these colours, even when you borrow from the natural world itself: apricot, lime, rose, lemon, carmine. Multi-petalled flowers, some flat as an outstretched palm, big as a basketball, small as a plum.

I am a model of restraint. I buy just one. My Beloved digs a hole without complaint, as he has done so often, our property at The Leaf now perforated with the holes he has dug. "I could be in China by now," he mutters, but he digs on just the same, twice as wide and just as deep as the pot that holds the plant. When we arrived here, there was a thin strip of garden beside the hedge at the front of the old stone house, a few roses where the limestone wall now curves around the terrace. I started small—some foundation beds, a perennial border, the Woodland Garden, the Winter Garden—until now, somehow, there are twenty-six.

"Isn't that enough?" he says as each one is finished.

"I'm not sure," I reply. "How much is enough?"

I set the rhododendron on the north side of the house,

where it is exposed to not a single deadly ray of winter sun. Every fall, I surround its puny form with a palisade of sumac branches wound with burlap from our local coffee-roasting merchant and stuffed with dry leaves that I rake from the grass. The Japanese maple and witch hazel have long since turned to compost, but the rhodo hangs on, its leathery leaves brightening when I unwrap it to the spring sun. For the first time last summer it set blooms, seven years after it dropped the lilac flowers it wore when I brought it home.

Last year I bought a magnolia. I was lusting after a cup-and-saucer magnolia, *Magnolia* × *soulangeana*, also known as the tulip tree. I had in mind a twenty-foot tulip tree like the one that overlooks the patio of the restaurant where my Beloved and I were married, the one under which we sit every year, from opening to closing time, marking that anniversary in the company of a succession of friends. The magnolia was named for Pierre Magnol, a French botanist who, before Linnaeus, classified plants according to their common features into families much like the Linnaean genealogy I struggle to remember now. Magnoliaceae is its own family, with some 250 species native to China, the Himalayas, Mexico, and Virginia. I like the fact that this is a very old plant, ninety-five million years at least. It is older than bees. The tough carpels that hold the stigma up to catch the pollen were designed to withstand the mauling of giant beetles. Dinosaurs nibbled on magnolias.

But is the magnolia tough enough for my garden? Yes, I was assured by the eager teenager at the instant nursery: this magnolia would certainly bloom for me. Its name is 'Royal Star' and it has narrow, strappy blooms, not the fat cups I love, but they are elegant in their own way and deliciously fragrant. *Magnolia stellata*. The tulip tree, alas, would never make it at The Leaf.

This one, though, just might stand a chance. It sets its blooms in the fall so they come on early in the spring, before the leaves.

That's what worries me, as I look out over the snow to the twigs of my fledgling star-magnolia. My Beloved and I planted it safely out of the winter wind, and out of the winter sun, too, so that its black bark won't heat up on a sunny day like this, tempting the sap to rise into the stems, where it can swell and freeze, cracking open the bark like a boil lanced too soon.

But how can I protect it from this cold? I might nurture it for years through statistically average winters, only to lose it in one single record-breaking day. For a tender shrub like the magnolia, once is enough. The buds that looked so lush in November, large, grey hairy pods like milkweed standing erect at the tip of every branch, look dangerously shrunken to me now. Wizened.

Will I lose it?

Perhaps.

Probably.

Will that stop me from forgetting myself when spring finally comes and I wander once again into nursery Never Never Land?

I doubt it. Some things can never be learned—and I'm not sure I want to.

INSPIRATION

I COLLECT GARDENS. It is an innocent enough hobby. They take up less space than china teacups or antique ink bottles, and they require neither dusting nor special insurance. They nestle unobtrusively at the back of my mind: gardens I've come across in books; gardens that friends have described to me; gardens I've seen and those I have yet to visit; gardens I can only imagine.

When I think of my collection, which I do on the first day of spring as I sit down with my papers and coloured pencils to consider my own few square feet of soil, it is the botanical gardens that tumble out first, acres of sprawling plantings labelled with brass plaques. They aren't all huge. Some, like the Chelsea Physic Garden, tucked in a corner of downtown London, England, are not much bigger than my gardens at The Leaf.

Botanical gardens began as physic gardens. The first one was established in Italy in 1544 by the botanist Luca Ghini, at the University of Pisa. The Chelsea Physic Garden was founded a hundred years later by the Worshipful Society of Apothecaries. Then, the study of plants was in the interest of medicine, and this garden keeps that bond alive, but what I kept for my collection was something else.

"We should get bees," my Beloved said as we left that garden.

"Yes!" We'd seen hives spotted about the "Mediterranean woodland," and we'd bought a pot of honey in the gift store. "We'll get some of those coiled straw hives. I'll plant a field of lavender."

Our eyes glazed with delight.

During the age of exploration, botany cut its ties with medicine and became a study unto itself. The plethora of plant material pouring into Europe from South America, Africa, and the East sparked our peculiar northern penchant for order. We developed systems of naming and classification, put the plants in "beds" and labelled them neatly, caging the exotics like animals in a zoo, simulating in these grand gardens the colonies that we managed with imperial control. Toward the end of the twentieth century, the role of botanical gardens shifted again, as we understood how little influence we have over living things. From symbols of imperial power, botanical gardens evolved into treasured reservoirs of horticultural diversity, public places to study the botanical ebb and flow.

We understood how little influence we have over living things.

When I visited the Chelsea Physic Garden in the 1970s—it was the first garden in my collection—there were eleven lichens on site. Who counts lichens? I wondered. Who cares? It seemed such an insignificant number, such an odd thing to do. But the last time I visited, I noticed the lichen count was up to thirty-eight and I was into the game. I cheered. Acid rain had been killing off the lichens, and now they were coming back. *Lecanora conizaeoides*, known as the "pollution lichen" for its dependence on sulphur dioxide, is now almost completely gone from the garden. As many as a dozen lichens graced a single park bench.

"Let's get a wooden park bench," I say. "Set it in the woods."

"It will rot," my Beloved says sensibly.

"I know. But we could count the lichen."

There are almost two thousand botanical gardens around the world. I have seen only a few. The rose garden in Paris, where I cupped my hand around a yellow rose about to let loose its petals and bent my nose to it, as the Rosarian so often does, provoking the wrath of a passing Frenchman. The Shakespeare Garden in a corner of the Bois de Boulogne, where five small gardens are planted with species mentioned in his plays. The Boboli Gardens behind the Pitti Palace in Florence, Italy, with their wide gravel avenues, exquisitely positioned sculptures, expansive views, and shimmering pools, though what I've stored in my mental collection is the maze of covered *allées* through the woods, the living fences made from saplings bent and woven together as they grow.

"How about a pond with a sculpture in the middle, like at the Boboli?"

It is March, the first day of spring. My horticultural blood is gushing.

My Beloved is kind. "That would be nice, but we don't have water," he reminds me gently.

"Neither did the Medicis." I eye the stream across the road. "They piped in water from the Arno River."

He raises an eyebrow. The stream feeds the neighbours' cows. It waters the Rosarian's roses. He doesn't want to start a Milagro Beanfield War. But I don't give up easily. "We could have lily pads. Giant lily pads. One blossom would scent the whole yard."

I've just come back from Rio de Janeiro, with a fresh specimen for my garden collection. The Jardim Botânico sits under the outstretched arms of Christ the Redeemer, as does the entire city, though it seems, looking up from the gardens, that he casts a protective eye in particular over those 346 acres that the king of Portugal planted with the spice trees he was so keen on introducing to Brazil. The great avenue leading into the park is planted with 134 palms, each one descended from a single tree, the *Palma Mater*, but the park itself contains almost a thousand other varieties of palms and six thousand species of tropical and subtropical plants, including a pond that seems tiled with vast, round Victoria lilies, huge-lipped trays big enough to support a small child.

I want a *grand allée* of palms. I want a water-lily garden. I want fountains and vistas and a garden planted with every species mentioned in Catharine Parr Traill's writing, or maybe Margaret Atwood's. I want a Pleasure-Garden. Why not? It's spring: everything is possible.

"An *allée*?" my Beloved says, squinting out at the lawn still brown with winter. "How grand?"

I'm sketching madly. "From the apple tree to the rail fence? And a pond there, in the middle? Water lilies: what do you do with them in winter? I wonder. And palms. Would they be happy here by the fire, do you think, for six months of the year?"

He wisely leaves me to my musings. He's been through this before. The grandiose gardens will blow through like great winged creatures released by the scent of spring, and when they've passed, the other gardens of my inspiration will tiptoe out. The lane I walked down in Serra Negra, where impatiens grew wild in the woods, not in garish mounds, but a plant here and there poking a single red, coral, or pink flower four feet up through the tangle of underbrush. Or the garden glimpsed from a hotel room, where a row of bamboo stood upright between a bamboo channel, support and plant seamlessly integrated. Or the brooms the street sweepers made for themselves by tying twigs to a pole, the spread of the branches precisely matched to the breadth of the path. It's the grace of that breadth I collect, the sweep of that pole, the tambourine shake of branch against the earth. And the canna lilies glimpsed growing wild by the roadside, palms waving at their backs. I collect those, too.

My pencil has slowed. The big arcs that would remodel the slope of The Leaf into an Italianate vista or a Brazilian plantation have spiralled down to small thumbnails, intimate clusters of companionable plants.

"I think maybe I'll put some canna lilies in front of the sumac this year," I say, looking out to where the feathery fronds will soon appear. "They look a lot like palms, don't you think?"

ACTS OF LOVE

I PUT THE GARDENS TO BED PROPERLY THIS YEAR. I say *this year,* because even though we are almost a third of the way into the new calendar year, until I am out there, my hands baptized in soil, the gardening year has not yet begun. For a few more days, I can revel in the satisfactions of a cycle completed before the new one thrusts me forward into the unknown. I feel the way I used to in the last days of my pregnancies: strangely calm, knowing I have done all I can, but alive with excitement, too, aware of what is ahead of me, uncertain that I will measure up, even after I have proved I can do this, have done it so often before.

The last snow recedes slowly, revealing crisp edges, the soft chocolate mounds of leaves. I watch the slow unveiling from my kitchen table, where my tools and honing stones, my scraps of flannelette from last year's nighties are spread before me. My hands are clumsy with lack of use. It takes me a while to get into the rhythm, the slick-slap-slick of metal against stone. By the end of April, I'll flick a razor's edge on these secateurs in seconds. I feel stupid having to relearn these small lessons every year, but maybe that's one of the things a garden teaches, that the growing never ends.

Making a garden is not that different from writing—each

book a fresh expanse of unworked soil, a mashup of ideas and images with a couple of characters and maybe a scene or two sitting there fully formed like shrubs bought at a nursery or dug from the woods. They are the bones of the story. The rest evolves as it will. I move the bits around endlessly until something about the arrangement seems pleasing and fitting within the overall landscape.

> Maybe that's one of the things a garden teaches, that the growing never ends.

I know a few things about sentence structure and punctuation. I know a few things about what a rose needs to bloom and when to prune a tree. I've learned mostly from the people I admire—the Rosarian, my Garden Guru, the Frisian, my Beloved. Learned to let the beds lie a little longer in the spring, give the volunteers a chance to sprout; learned not to worry so much, to watch instead and see what the garden is telling me in the colour of its leaves, the shape of its branches, the tangle of its roots; learned to dig with gusto and optimism; learned to give it a rest.

I slip the freshly sharpened secateurs into their leather holster and slide them, together with the holstered gardening knife, onto the belt my Beloved gave me for Christmas. With luck, I'll never lose my tools again.

Have I figured out this urge to plant my hands in the earth? Not really. I would like to think it's something noble, or primitive, an urge to be part of the great living planet that in our blind wisdom we call Earth. But I think it's just as likely that it is grounded in a less honourable, more problematic trait—that awful human drive to move and manipulate, to make a mark, to claim what we see and want as ours.

"Are you coming outside?" my Beloved says, poking his head into my study. "The sun is shining. It looks warm."

And so I close my computer and follow him down the stairs.

There's something in a garden that resists theories, even ideas as rudimentary as mine. I brush aside the leaves and bend to the stirring plants. *Give it up,* I hear them laughing. *Come and play.*

My Beloved takes my hand as we walk among the beds, and I think, There is one thing, at least, that I've learned—that a garden, like love, is an act of imagination and faith.

ACKNOWLEDGMENTS

CAN A PLANT BE THANKED? What's the point, you might say, but even so, it is the plants I want to acknowledge, for their endless inspiration, the joy they bring, and the frustrations that make me question what I do and why.

I am grateful, too, for the particular landscape of The Leaf and its custodians—the Moltons, the Heffernans, Anna Bentley, Gwen and Hugh Stewart.

My Beloved; my neighbours; the Rosarian and the Humanist; my great helper, the Frisian; and my good and loyal friend the Garden Guru: I cannot imagine growing without you.

I write within a tradition of writer/gardeners who have dug in the earth and contemplated what they found there— Colette, Karel Čapek, Germaine Greer, Gertrude Jekyll, Jamaica Kincaid, Patrick Lane, Elizabeth Lawrence, Eleanor Perenyi, Catharine Parr Traill, Vita Sackville-West, Elizabeth Smart, to name but a few. These writers confirm for me the essential significance of the garden, not only as a turning point in human history, the moment when we first put down roots, but as the threshold between nature and human nature.

In 2009, in a fit of despair at being unable to garden for a time, I launched a website called the Frugalista Gardener, where

I posted many of these essays over the course of a year. The response of readers and gardeners renewed my faith in the ephemeral. One of those readers was Nita Pronovost, an editor who shares my love of words and gardens. My deep appreciation to her for a sensitive, respectful edit, and to the Latta sisters, Allyson and Lenore, for their careful copyedit, which made of me a more precise gardener than I often am. And a special thank you to the book's designer, Leah Springate.

The Grand Girls, my Elder and Younger Sons, my Beloved's Daughters, my Friend in the North, the Poet, the Carpenter, my sisters, my friend Linda at Maple Lane and Craig at Wheatley Wood, my friends, and all the others, mentioned in these pages or not, I thank you. You are welcome in my garden any time.

Finally, to my agent, Bella Pomer, a bouquet of hyacinthe, rose, and forget-me-not, for her constancy, love, and loyal support.